Synthesis Lectures on Tissue Engineering

This series publishes concise books on aspects of a field that holds so much promise for providing solutions to some of the most difficult problems of tissue repair, healing, and regeneration. The field of tissue engineering straddles biology, medicine, and engineering, and it is this multi-disciplinary nature that is bound to revolutionize treatments for a plethora of tissue and organ problems. Central to tissue engineering is the use of living cells with a variety of biochemical or biophysical stimuli to alter or maximize cellular functions and responses. However, in addition to its therapeutic potential, this field is making significant strides in providing diagnostic tools. Topics will include: 1) tissue engineering knowledge on particular tissues or organs (e.g., articular cartilage, liver, cardiovascular tissue), but also on 2) methodologies and protocols, as well as 3) the main actors in tissue engineering paradigms, such as cells, biomolecules, biomaterials, biomechanics, and engineering design.

Rachel C. Nordberg · Takumi Takahashi ·
Jerry C. Hu · Kyriacos A. Athanasiou

Tissue Engineering the Cartilages of the Body

Hyaline Cartilage, Fibrocartilage, Elastic Cartilage, and Lipocartilage

Springer

Rachel C. Nordberg
Department of Biomedical Engineering
University of California, Irvine
Irvine, CA, USA

Jerry C. Hu
Department of Biomedical Engineering
University of California, Irvine
Irvine, CA, USA

Takumi Takahashi
Department of Biomedical Engineering
University of California, Irvine
Irvine, CA, USA

Kyriacos A. Athanasiou
Department of Biomedical Engineering
University of California, Irvine
Irvine, CA, USA

ISSN 1944-0316 ISSN 1944-0308 (electronic)
Synthesis Lectures on Tissue Engineering
ISBN 978-3-032-09104-8 ISBN 978-3-032-09105-5 (eBook)
https://doi.org/10.1007/978-3-032-09105-5

This Springer imprint is published by the registered company Springer Nature Switzerland AG
The registered company address is: Gewerbestrasse 11, 6330 Cham, Switzerland

To my mother, Beth, and father, John, for their belief in me and for raising me to pursue my dreams. Their love and support have been integral every step of the way.—Rachel C. Nordberg

To my parents, Kayoko and Takashi, whose unwavering support and encouragement have guided me throughout every pursuit. To my wife, Yoshiko, and my son, Kai, whose love and joy bring meaning to both my life and my work.—Takumi Takahashi

To family, friends, and our past and current students, may we continue to find inspiration and energy in our pursuits.—Jerry C. Hu

For my grandchildren, Theodore and Mary Athens—may your lives be filled with curiosity, discovery, and the joy of asking "why." May you always find wonder in the world around you and never stop learning.—Kyriacos A. Athanasiou

Preface

Cartilages, appreciated for their structural and functional roles in diverse anatomical locations, pose difficult problems to clinicians and scientists because of their limited intrinsic capacity for repair. While the regenerative limitations of hyaline cartilage have historically drawn the most attention, other cartilage types, fibrocartilage, elastic cartilage, and the newly recognized lipocartilage present equally significant needs and challenges in tissue engineering. From joint surfaces and temporomandibular discs to the airway, ear, and nose, the need to understand and to engineer functional cartilage tissues across the various types of cartilages has become increasingly evident.

The field of cartilage tissue engineering has grown rapidly since the formal establishment of tissue engineering as a discipline in the 1990s. Advances in stem cell biology, biomaterials, biomechanics, single-cell and spatial omics, and preclinical animal models have propelled the field toward clinical translation. However, despite decades of research, the number of approved tissue-engineered cartilage products available on the market remains limited.

This book provides a comprehensive and current overview of the primary cartilage types and the latest strategies for their tissue engineering. It begins with an overview of general cartilage types and detailed discussion on select cartilages located throughout the body. Subsequent chapters examine recent approaches for tissue engineering each cartilage type, including reviews on cell sources, scaffold-based and scaffold-free systems, and biochemical and biomechanical stimulation methods. Methods for evaluating native and engineered cartilage are presented alongside key regulatory and translational considerations for advancing therapies from bench to clinic.

This text is intended for undergraduate and graduate students, academic researchers, clinicians, and industry professionals engaged in or entering the field of cartilage tissue engineering. As discoveries continue to accelerate, we hope this serves as both a foundational reference and a springboard for future innovation in the tissue engineering of cartilages.

Irvine, CA, USA

Rachel C. Nordberg
Takumi Takahashi
Jerry C. Hu
Kyriacos A. Athanasiou

Contents

About the Authors

Rachel C. Nordberg (co-first author) is a research associate at the University of California, Irvine. Her research focus is on the translation of cell therapy products with a particular emphasis on musculoskeletal tissue engineering. She earned her B.S. in Bioengineering from Lehigh University before attending Cornell University for her M.Eng. in Biomedical Engineering. She obtained her Ph.D. in Biomedical Engineering from the University of North Carolina, Chapel Hill and North Carolina State Joint Department of Biomedical Engineering. In 2018, she began her postdoctoral research at the University of California, Irvine where she was awarded a three-year TL-1 NRSA Postdoctoral Fellowship from the UCI Institute of Clinical and Translational Science.

Takumi Takahashi (co-first author) is a postdoctoral fellow at the University of California, Irvine. He earned his B.S. in Bioengineering from the University of California, San Diego and his Ph.D. in Medicine from the Department of Orthopaedic Surgery at Tokai University, Japan. In his postgraduate and postdoctoral work, his research has focused on translating scaffold-less approaches to cartilage tissue engineering for the treatment of musculoskeletal diseases such as osteoarthritis. He has experience in first-in-human clinical studies, conducted for the autologous and allogeneic transplantation of chondrocyte cell-sheets, and in preclinical studies using various small and large animal models.

Jerry C. Hu is a principal development engineer at the University of California, Irvine. He obtained a PhD in Bioengineering from Rice University and a BSc degree in Chemical Engineering from the University of Texas at Austin. Dr Hu has over 15 years of experience in the study and development of materials, bioreactors, and culture methods for articular cartilage tissue engineering. Jerry has made significant contributions to the research of developing new approaches for cartilage tissue engineering, such as the formation of neotissue without a scaffold, the design and application of mechanical stimulators, and the acquisition of novel autologous cell sources. His direction and collaboration have led to numerous publications, presentations, and other accolades, including the filing of three patents as co-inventor.

Kyriacos A. Athanasiou (corresponding author) has addressed significant societal needs through the development of life-saving technologies. He is a Distinguished Professor at the University of California, Irvine. He is also the holder of the Henry Samueli Chair and the Director of DELTAi (Driving Engineering and Lifescience Translational Advances at Irvine). He has served as a faculty member at the University of Texas, Rice University, and the University of California (first at Davis and now at Irvine). Prof. Athanasiou obtained his PhD in Bioengineering (Mechanical Engineering) from Columbia University in 1989. He has published over 800 peer-reviewed publications, including 400 papers, 370 conference proceedings or abstracts, 20 books, and multiple patents. He has also served as President of the Biomedical Engineering Society, and as the Editor-in-Chief of the Annals of Biomedical Engineering and has received numerous awards. He was elected to the National Academy of Engineering in 2024, the National Academy of Medicine in 2020, and the National Academy of Inventors in 2014.

Abbreviations

3D	Three-dimensional
ACI	Autologous chondrocyte implantation
ACL	Anterior cruciate ligament
ACP	Analogous cellular product
AD-MSC	Adipose-derived mesenchymal stromal/stem cell
AFM	Atomic force microscopy
AGE	Advanced glycation end-product
ASTM	American Society for Testing and Materials
BBB	Basso, Beattie, and Bresnahan
BLA	Biologics License Application
BLOKS	Boston-Leeds Osteoarthritis Knee Score
BMAC	Bone marrow aspirate concentrate
BM-MSC	Bone marrow-derived mesenchymal stromal/stem cell
BMP	Bone morphogenetic protein
BODIPY	Boron-dipyrromethene
c-ABC	Chondroitinase-ABC
CBC	Complete blood count
CBER	Center for Biologics Evaluation and Research
CDER	Center for Drug Evaluation and Research
CDRH	Center for Devices and Radiological Health
CFR	Code of Federal Regulations
cGMP	Current Good Manufacturing Practice
cGTP	Current Good Tissue Practices
CMC	Chemistry, manufacturing, and control
CMI	Collagen Meniscus Implant
COMP	Cartilage oligomeric matrix protein
CPC	Cartilage-derived progenitor cell
CRO	Contract research organization

CT	Computed tomography
DMMB	1,9-dimethyl methylene blue
DNA	Deoxyribonucleic acid
ECM	Extracellular matrix
EDTA	Ethylenediaminetetraacetic acid
ESC	Embryonic stem cell
EV	Extracellular vesicle
FAA	Fibronectin adhesion assay
FACS	Fluorescence-activated cell sorting
FBS	Fetal bovine serum
FDA	United States Food and Drug Administration
FGF	Fibroblast growth factor
FI	Functionality index
FLIM	Fluorescence lifetime imaging microscopy
GAG	Glycosaminoglycan
GDF	Growth differentiation factor
GLP	Good laboratory practice
H&E	Hematoxylin and eosin
HA	Hyaluronic acid
HCT/P	Human cells, tissues, and cellular and tissue-based product
HIF-1α	Hypoxia-inducible factor 1α
ICH	International Council for Harmonisation
ICRS	International Cartilage Regeneration and Joint Preservation Society
ICRS II	ICRS Visual Assessment Scale II
IDCT	Injectable disc cell therapy
IGF	Insulin-like growth factor
IHC	Immunohistochemistry
IMS	Imaging mass spectrometry
IND	Investigational New Drug
INTERACT	INitial Targeted Engagement for Regulatory Advice on CBER/CDER ProducTs
IP	Intellectual property
IPO	Initial public offering
iPSC	Induced pluripotent stem cell
ISO	International Organization for Standardization
ITS	Insulin-transferrin-selenium
IVD	Intervertebral disc
KOSS	Knee Osteoarthritis Scoring System
LC-MS/MS	Liquid chromatography-tandem mass spectrometry
LLC	Lower lateral cartilage
LOXL2	Lysyl oxidase-like 2

MACI	Matrix-induced autologous chondrocyte implantation
MALDI	Matrix-assisted laser desorption/ionization
m-AMI	Matrix-induced autologous BM-MSC implantation
MCB	Master cell bank
MOAKS	MRI Osteoarthritis Knee Score
MOCART	Magnetic Resonance Observation of Cartilage Repair Tissue
MODS	Modified O'Driscoll Scale
MRI	Magnetic resonance imaging
MSC	Mesenchymal stromal/stem cell
N-cadherin	Neural-cadherin
N-CAM	Neural cell adhesion molecule
NIH	National Institutes of Health
OA	Osteoarthritis
OARSI	Osteoarthritis Research Society International
PCL	Posterior cruciate ligament
PDGF	Platelet-derived growth factor
PHS	Public Health Service
Picrosirius red	Sirius red in picric acid
PRG4	Proteoglycan-4
PRP	Platelet-rich plasma
PTHrP	Parathyroid hormone-related protein
PTIBS	Porcine Thoracic Injury Behavior Scale
QMS	Quality Management System
RA	Rheumatoid arthritis
RMAT	Regenerative Medicine Advanced Therapy
RNA	Ribonucleic acid
SBIR	Small Business Innovation Research
scRNA-seq	Single-cell RNA-sequencing
SEM	Scanning electron microscopy
SOP	Standard operating procedure
SOX9	SRY-box transcription factor 9
STTR	Small Business Technology Transfer
SY-MSC	Synovium-derived mesenchymal stromal/stem cell
TEM	Transmission electron microscopy
TGF-β	Transforming growth factor-beta
TIBS	Thoracic Injury Behavior Scale
TMD	Temporomandibular disorder
TMJ	Temporomandibular joint
TOF-SIMS	Time-of-Flight Secondary Ion Mass Spectrometry
TTO	Technology transfer office
UC-MSC	Umbilical cord-derived mesenchymal stromal/stem cell

ULC	Upper lateral cartilage
USPTO	United States Patent and Trademark Office
UTS	Ultimate tensile strength
VC	Venture capitalist
VEGF	Vascular endothelial growth factor
WCB	Working cell bank
Wnt	Wingless-type MMTV integration site family member
WORMS	Whole-Organ Magnetic Resonance Imaging Score

Cartilage Types, Development, and Injury/Degeneration

1

Takumi Takahashi, Rachel C. Nordberg, Jerry C. Hu, and Kyriacos A. Athanasiou

Cartilages are ubiquitous throughout the human body [1]. They are specialized connective tissues that fulfill distinct functions depending on their anatomical location. The hallmarks of cartilages are their dense extracellular matrix (ECM) and relatively low cellularity compared to most other tissues. Cartilages are traditionally categorized into three major types: hyaline cartilage, fibrocartilage, and elastic cartilage, based on the presence and relative abundance of collagens (primarily types I and II) and elastin (Fig. 1.1). Hyaline cartilage, the most prevalent type, covers the articulating surfaces of bones in synovial joints such as the knee, hip, and shoulder, facilitating low-friction movement and efficient load distribution [2, 3]. It also contributes to the structural integrity of the nose, trachea, bronchi, and costal cartilages of the rib cage, where it helps maintain airway patency and allows for thoracic expansion during respiration [2, 4, 5]. Fibrocartilage is located, for example, in the intervertebral discs (IVDs), pubic symphysis, knee menisci, and temporomandibular joint (TMJ) discs, where it bears and distributes loads by means of its dense collagen fiber network [6]. Elastic cartilage, noted for its flexibility, provides structural support in the external ear, epiglottis, and certain parts of the larynx [2, 7, 8]. More recently, a fourth type, termed lipocartilage, has been identified [9]. This fat-containing cartilage, previously overlooked in classical anatomy, is found in specialized regions such as the tip of the nose and parts of the ear and larynx of some mammalian species. Each cartilage type is unique in its biochemical composition, structural organization, and set of mechanical functions, posing distinct challenges for tissue engineering strategies aimed at repair, replacement, or regeneration. This chapter introduces the defining characteristics of hyaline cartilage, fibrocartilage, elastic cartilage, and lipocartilage, with an emphasis on their biochemical makeup, structural features, and functional or biomechanical roles. It then provides

R. C. Nordberg et al., *Tissue Engineering the Cartilages of the Body*, Synthesis Lectures on Tissue Engineering, https://doi.org/10.1007/978-3-032-09105-5_1

an overview of their developmental biology and common modes of injury or degeneration, laying the groundwork for subsequent chapters addressing anatomical variations, engineering approaches, and translational considerations in cartilage tissue engineering.

1.1 Hyaline Cartilage

Hyaline cartilage, derived from the Greek hyalos (ὕαλος), meaning "glassy" or "transparent," is the most prevalent and extensively characterized type of cartilage in the human body [2]. As its etymology suggests, freshly isolated hyaline cartilage exhibits a smooth, translucent appearance reminiscent of glass, a feature that distinguishes it from the other cartilage types. Hyaline cartilage is widely distributed [11]. It caps the articular surfaces of bones in synovial joints, forms the costal cartilages (connecting ribs to the sternum), makes up the structural cartilage of the nose, and supports the walls of the larger respiratory passages (trachea, bronchi, and some laryngeal cartilages). The body has three types of joints known as synarthrosis, amphiarthrosis, and diarthrosis, which are classified according to the degree of movement they permit [12]. A synarthrodial joint has almost no range of motion, such as the coronal suture of the skull. An amphiarthrodial joint, such as an intervertebral joint, allows for a limited range of motion. However, most joints that support locomotion in the body are classified as diarthrodial joints, which are characterized by having a large range of motion. In diarthrodial joints, hyaline cartilage acts to transmit loads from bone to bone and provides a lubricating surface on which bones can smoothly articulate. There are numerous diarthrodial joints in the body including the knee, hip, facet, phalangeal, temporomandibular, wrist, and ankle joints, and many more. Hyaline cartilage also serves as the template for endochondral ossification during skeletal development [13]. For example, the embryonic long bones are first laid down as hyaline cartilage, and hyaline cartilage persists at the epiphyseal growth plates of growing bones. In adults, hyaline cartilage remains as the articular cartilage capping the long bones and in permanent cartilaginous structures like the nasal septum and certain respiratory cartilages. Functionally, hyaline cartilage is a resilient, smooth, load-bearing tissue with high resistance to compression but relatively weak resistance to shear or tension [14]. Its composition, organization, and function have been the most studied among cartilages and, therefore, provides a baseline for the understanding of the other cartilage types.

1.1.1 Biochemical Composition of Hyaline Cartilage

The ECM of hyaline cartilage is rich in type II collagen and proteoglycans, which, together, form a highly hydrated, gel-like network. By weight, water is the single largest component, comprising roughly 65–80% of the total tissue weight [15]. This water, attracted by the proteoglycans, is held within the matrix and provides resilience, the ability

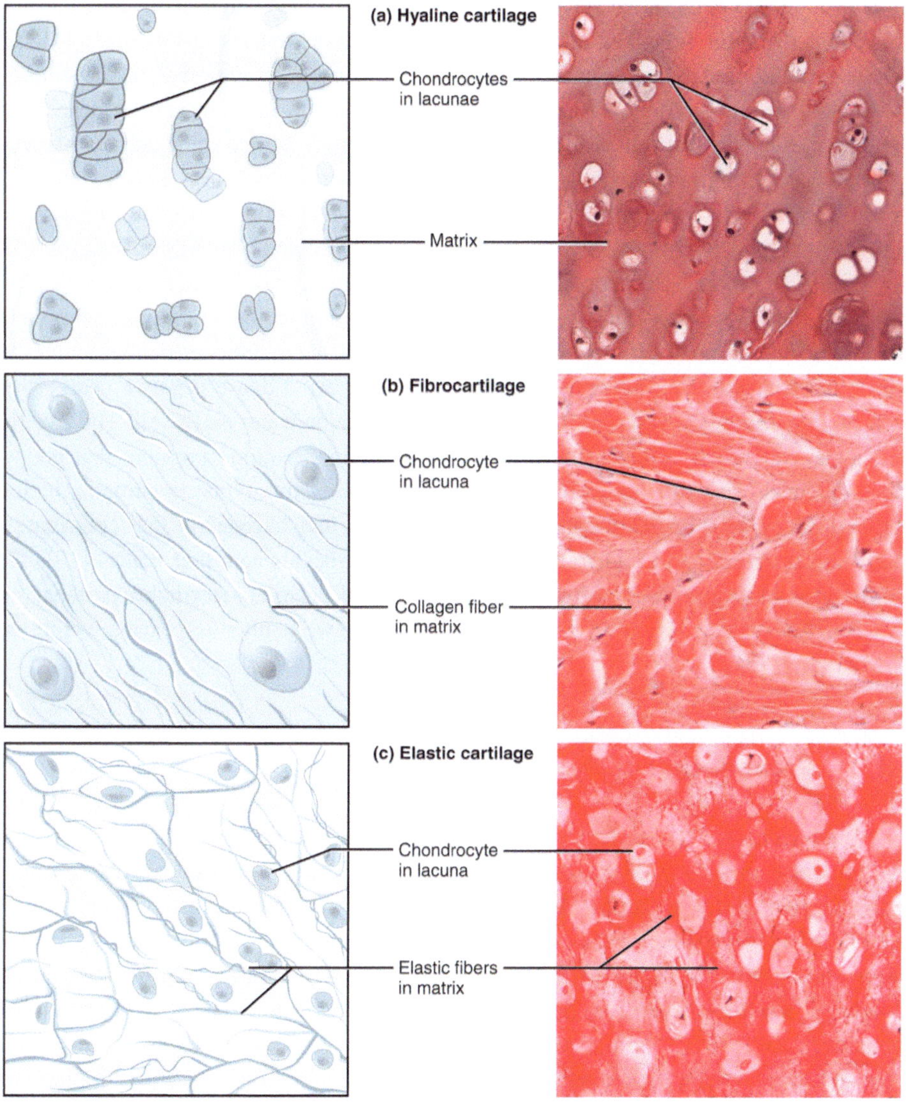

Fig. 1.1 Classical cartilage types. Cartilages are specialized connective tissues composed of collagen fibers embedded within dense extracellular matrices (ECMs). **a** Hyaline cartilage features a collagen-rich matrix with chondrocytes arranged in a columnar organization in the deep zone; the image shown is from canine tissue. **b** In fibrocartilage, chondrocytes are dispersed within a dense fibrous collagen matrix. **c** Elastic cartilage contains chondrocytes embedded in a matrix rich in elastin fibers. Reproduced with permission [10]

to deform and recover, to cartilage. The vast majority of dry weight of hyaline cartilage is accounted for by collagens at roughly 60% and proteoglycans at 25–30% [11]. The collagen component is predominantly type II collagen fibrils (90–95% of the collagen in hyaline cartilage), which are thin (~20 nm diameter) and form an interlacing network that provides tensile strength while still allowing flexibility. Minor collagens, such as type IX and type XI, are interwoven at small percentages (each < 5% of collagen content) to stabilize the fibrillar network [11]. Type IX collagen, for example, is a fibril-associated collagen that links type II collagen fibrils to proteoglycans, and type XI collagen regulates fibril size. The proteoglycan content is dominated by aggrecan, a large chondroitin sulfate/keratan sulfate proteoglycan [16]. Hundreds of aggrecan monomers bind via link proteins to long hyaluronic acid (HA) chains, forming proteoglycan aggregates that are enmeshed within the collagen fibril network. These aggregates carry a dense negative charge from their glycosaminoglycans (GAGs), causing them to imbibe water and swell. The swelling is constrained by the collagen mesh, creating substantial osmotic pressure that endows the tissue with resistance to compression [17]. Notable minor matrix proteins include cartilage oligomeric matrix protein (COMP), decorin, and fibronectin that contribute to matrix organization and to interactions between chondrocytes and their matrix [18]. This composition of a tight weave of collagen fibrils embedded in a water-laden proteoglycan gel gives hyaline cartilage a firm, viscoelastic structure well-suited to bearing weight and distributing loads.

1.1.2 Organization of Hyaline Cartilage

In hyaline cartilage, chondrocytes, the cells found in cartilage, make up only ~ 1–5% of the tissue volume [19]. They are typically solitary and sparsely distributed, each occupying a small cavity in the ECM called a lacuna (Fig. 1.2). During development or in juvenile cartilage, chondrocytes can divide and form small clusters of daughter cells of two to four cells in one lacuna that are called isogenous groups, indicating recent cell division. Chondrocytes are round and are surrounded by a thin specialized matrix layer, the pericellular matrix. The pericellular matrix is rich in type VI collagen and proteoglycans and forms a capsule around the cell (termed the chondron unit) [20]. Immediately around each lacuna is the territorial matrix, a narrow zone with a higher concentration of fine collagen fibrils (including collagen VI and IX) and lower aggrecan content, weaving around the chondrocyte or cell cluster. Beyond that lies the interterritorial matrix, which is the bulk of the ECM filling the space of cartilage. Here, the collagen fibrils (predominantly type II collagen) are thicker and more densely packed, and this is where most of the large proteoglycan aggregates reside. This gradation from pericellular to territorial to interterritorial matrices protects chondrocytes from mechanical forces while transmitting important mechanical signals [21]. In short, the pericellular matrix facilitates mechanotransduction to the chondrocyte, whereas the interterritorial matrix provides a strong load-bearing framework.

A hallmark of hyaline cartilage serving as articular surfaces in synovial joints is its depth-wise zones (Fig. 1.3). From the articulating surface down to the subchondral bone, articular cartilage is organized into discernible zones, reflecting functional adaptation to

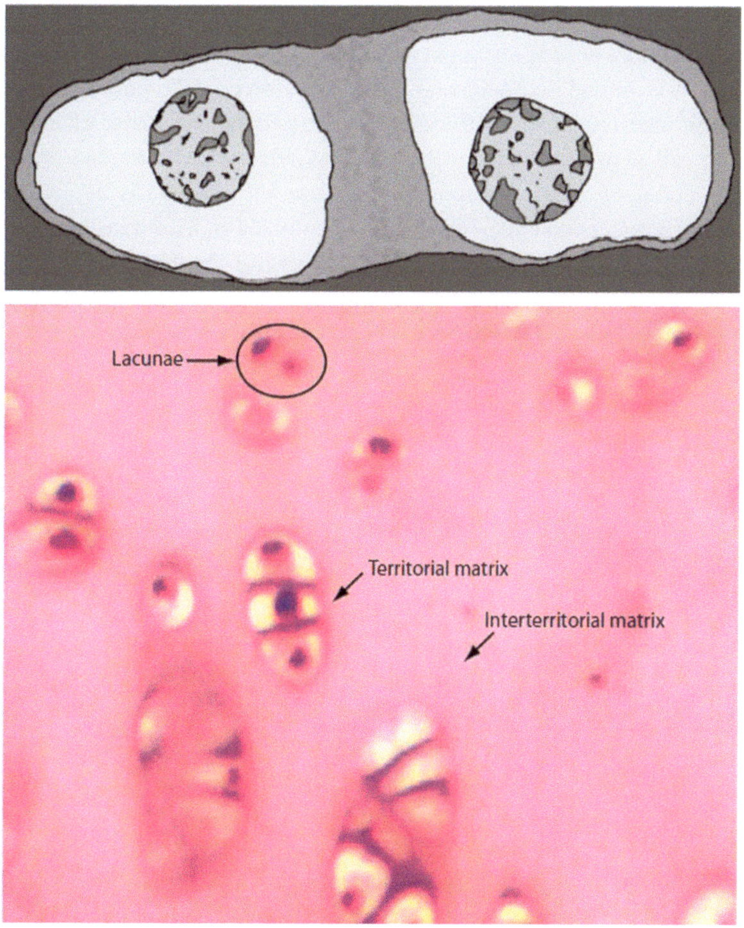

Fig. 1.2 Structure of the chondrocyte microenvironment. Top: Chondrocytes (light gray) are housed within chondrons (medium gray), each surrounded by a territorial matrix (dark gray). Beyond this zone lies the interterritorial matrix, which comprises the bulk of the cartilage tissue. Together, the territorial and interterritorial regions constitute the extracellular matrix. Bottom: Histological image illustrating the ECM architecture surrounding chondrocytes. Adapted [22] and reproduced [2] with permission

gradients of stress [23]. In the superficial (tangential) zone (top ~ 10–20% of thickness), chondrocytes are small and flattened and oriented parallel to the surface. Collagen fibrils in this zone are very fine and run parallel to the articular surface, forming a tight, woven mat. This structure provides a smooth, almost frictionless surface and high tensile strength to resist shear forces at the surface. The superficial zone also has the highest water content (~80%) and lowest proteoglycan content (~2–4%) [2, 24]. Importantly, superficial zone chondrocytes and cells that line the synovial joint are the source of lubricin (also known as proteoglycan-4, PRG4), a mucinous glycoprotein that is secreted into the joint and coats the articular cartilage surface to greatly reduce friction [25, 26]. Lubricin from these surface cells, together with HA from synovial fluid, forms a boundary lubricant film that protects against wear and cell apoptosis from shear forces [27]. Immediately below, the middle (transitional) zone (~40–60% of thickness) contains rounder chondrocytes that are lower in cell density. Collagen fibrils in this zone are thicker and arranged more randomly or obliquely. The proteoglycan concentration (~5–7%) is higher here than at the surface, while water content (~70–75%) is slightly lower. This transitional zone serves as an intermediate layer that functions as the first line of resistance against compressive forces from the surface, and it provides a buffer between the surface and deep regions [28]. Below, the deep zone (~20–50% of thickness) is characterized by large chondrocytes often aligned in columnar stacks perpendicular to the surface. The collagen fibrils in the deep zone are the thickest and are oriented vertically (radially), anchoring firmly into the calcified layer below. In the deep zone, proteoglycan content (~8–10%) is highest and water content (~65–70%) is lowest, conferring the greatest resistance to compressive forces [15]. Finally, the calcified cartilage zone is a thin region separating the cartilage from the subchondral bone, marked by an undulating boundary called the tidemark. In the calcified zone, the matrix is mineralized, impregnated with calcium salts, and the few chondrocytes present are hypertrophic and metabolically quiescent [29]. This calcified layer anchors the cartilage securely to the bone.

Notably, hyaline cartilage of articular surfaces (articular hyaline cartilage) lacks a perichondrium, the fibrous connective tissue sheath found in many other cartilages [2]. This absence is essential for enabling smooth cartilage-to-cartilage contact within the joint. Articular hyaline cartilage is nourished by diffusion from the synovial fluid, and the integrity of its superficial zone is critical for joint health. When the surface collagen network is compromised or the lubricating layer is disrupted, friction increases, accelerating cartilage wear and degeneration. In contrast, hyaline cartilage in non-joint locations (e.g., nasal septum, tracheal rings, and costal cartilage) is covered by a perichondrium [4, 5]. The perichondrium consists of an outer fibrous layer (dense collagenous tissue with fibroblasts) and an inner chondrogenic layer that harbors progenitor cells, referred to as chondroblasts, capable of appositional cartilage growth [30]. The perichondrium is vascularized and supplies nutrients to the underlying cartilage by diffusion and contributes to growth and repair. Overall, hyaline cartilage features an organized structure, including a collagen-proteoglycan network, depth-dependent zonal architecture, and, in

Fig. 1.3 In articular hyaline cartilage, both the cellular morphology (left) and collagen fiber orientation (middle) are organized into distinct depth-dependent zones, superficial, middle, and deep, which occupy approximately 10–20%, 40–60%, and 20–50% of the full cartilage depth, respectively. Safranin-O/fast green staining indicates the glycosaminoglycan content in the tissue (right). Reproduced with permission [2]

certain anatomical (non-articular) sites, a perichondrium that supports its maintenance and regenerative potential.

1.1.3 Functions of Hyaline Cartilage

The structure and organization of hyaline cartilage is especially suited to its functions in the body. Mechanically, it behaves as a firm but flexible material capable of withstanding compressive forces and providing a near-frictionless articulating surface [31]. In a joint, when compressive load is applied, water is squeezed out of the aggrecan-rich matrix over time, and the negatively charged GAG chains repel each other. This fluid pressurization and matrix stiffness create a hydrostatic cushioning effect. Once the load is removed, the tissue is gradually rehydrated, restoring its shape. Viscoelastic response allows articular hyaline cartilage to accommodate compressive forces during joint loading, such as the impact of running or jumping, which can impose in the knee as much as approximately 3.5 times the body weight in force [2]. In addition, the collagen network provides tensile strength to counteract swelling pressure from proteoglycan-driven water attraction [11]. However, because hyaline cartilage contains predominantly type II collagen fibrils that are not arranged in dense bundles, it is not as strong against shear or tension as fibrocartilage [32]. Hyaline cartilage is less resistant to tensile forces than fibrocartilage and typically exists in locations such as well-aligned joints where shear and tension are minimized.

Within a normal joint, the zonal organization mitigates these weaknesses. The surface lubrication by lubricin and HA gives articular cartilage an extremely low coefficient of friction (~0.001–0.03, lower than ice on ice, which is ~ 0.03–0.1), enabling smooth, wear-resistant motion of joints [33]. Another functional aspect of hyaline cartilage is its role in distribution of load. In a joint, the deformability of cartilage under load allows the distribution of forces over a larger contact area, reducing peak stresses on subchondral bone. In structures like the costal cartilages or tracheal rings, hyaline cartilage provides semi-rigid support that is stiff enough to maintain open airways or stable ribcage structure, yet flexible enough to allow breathing movements. For a more in-depth discussion of the mechanical properties of various hyaline cartilages, please refer to Chap. 2.

The metabolic turnover of hyaline cartilage is functionally significant. Because of its avascularity and low cellularity, hyaline cartilage has a very slow turnover and low oxygen demand. Chondrocytes rely on anaerobic, or hypoxic, metabolism and diffusion of nutrients, and, in healthy adult cartilage, they typically remain in a quiescent state maintaining the matrix [34]. This contributes to the longevity of the matrix, which can function for decades, but also limits its intrinsic repair capacity after damage. Minor injuries often do not heal, and larger defects often fill with scar-like fibrocartilage or not at all [35]. Thus, the composition and structure of hyaline cartilage endow it with unique load-bearing and low frictional properties essential for smooth joint articulation and load distribution but with minimal regenerative capacity. These characteristics underscore the degenerative nature of hyaline cartilage once damaged and the motivation for strategies to tissue-engineer hyaline-like cartilage.

1.2 Fibrocartilage

Fibrocartilage, derived from the Latin *fibra*, meaning "filament-like," is a tough, fibrous form of cartilage that represents a hybrid between dense connective tissue (tendon/ligament) and hyaline cartilage [2, 6]. Macroscopically, fibrocartilage appears white, opaque, and lacks the glassy translucency of hyaline cartilage [32]. It is found in regions of the body subjected to high mechanical stresses that involve both compression and tension. Examples include the IVDs of the spine with fibrocartilage found in the annulus fibrosus, the knee menisci, the pubic symphysis joining the two pubic bones, the TMJ discs, and certain entheses (tendon/ligament attachment sites to bone, such as the Achilles tendon insertion). For the latter, fibrocartilage often serves to transfer muscle forces from tendon to bone [36]. Unlike other cartilages, fibrocartilage usually lacks a distinct perichondrium because it often transitions gradually into the neighboring tissues, e.g., merging with ligament or bone. Functionally, fibrocartilage is specialized to resist strong tensile forces while still resisting compression [6]. It is the strongest cartilage in tension and shear, making it essential in structures that must not fail under pulling forces. For example, the menisci bear hoop stresses in the knee, and the annulus fibrosus of the spine contains the nucleus pulposus under pressure, and the TMJ disc serves as a trampoline resisting tensile forces [32]. Microscopically, fibrocartilage contains dense bundles of collagen fibers in the matrix with rows of relatively sparse chondrocytes aligned between the fibers. The combination of tensile strength and compressive resistance allows fibrocartilage to function as a tough, stress-bearing interface in high-load regions of the body.

1.2.1 Biochemical Composition of Fibrocartilage

The ECM of fibrocartilage is dominated by type I collagen, the same collagen type found in tendon and bone [11]. This contrasts with hyaline and elastic cartilages, which are rich in type II collagen. Fibrocartilage contains both type I and type II collagens, but type I collagen is predominant, especially in high-stress fibrocartilaginous regions. For example, in the annulus fibrosus of an IVD, studies indicate roughly 60–70% of the collagen is type I, with most of the remainder being type II, with more type II near the nucleus pulposus [37]. In the knee meniscus, a similar high type I to type II collagen ratio exists [11]. In the TMJ disc, the type I to type II collagen ratio is higher; it is primarily composed of type I collagen, with minimal type II content [38]. The proteoglycan content in fibrocartilage is low when compared to hyaline cartilage. Consequently, fibrocartilage's ability to attract water and generate swelling pressure is lower. Therefore, fibrocartilage resists compressive loads primarily through its dense and aligned collagen fiber network, with a lesser contribution from proteoglycan-induced swelling pressure compared to hyaline cartilage. Histologically, fibrocartilage stains acidophilic (pink/red with hematoxylin & eosin (H&E) staining) because of the abundance of type I collagen (eosinophilic) and paucity of GAGs (basophilic). Minor components of fibrocartilage may include small amounts of elastin fibers in certain fibrocartilages. For example, the outer rim of the knee meniscus and the pubic symphysis in late pregnancy can contain elastin fibers, but elastin is generally not a prominent feature. Fibrocartilage, in summary, is characterized by type I collagen and densely packed collagen bundles.

1.2.2 Organization of Fibrocartilage

Fibrocartilage, being an intermediate between tendon and cartilage, often contains two cell types with distinct phenotypes: chondrocytes and fibroblast-like cells, often referred to as fibrochondrocytes [39]. The chondrocytes in fibrocartilage are typically smaller and more ovoid (elongated) than the rounded chondrocytes of hyaline cartilage (Fig. 1.1) [38]. They reside in lacunae within the matrix, but the lacunae can be more irregularly shaped because of the surrounding collagen bundles. These chondrocytes produce cartilage matrix, including type I and type II collagens and proteoglycans, but they are often found in distinct arrangements. Type I collagen fibers in fibrocartilage are often organized into thick bundles or lamellar sheets aligned along principal stress directions. In structures such as the knee meniscus and annulus fibrosus, the collagen is arranged in concentric lamellae, forming a laminated structure that enables the tissue to withstand complex, multidirectional tensile forces [3]. For instance, in the knee meniscus, circumferential collagen bundles align with rows of chondrocytes, while in the annulus fibrosus, chondrocytes are found along the boundaries of successive lamellae [40]. These lamellae refer specifically to histologically distinct, layered collagen zones (not to be confused with bony lamellae) each oriented at alternating angles to enhance mechanical resistance. The fibrochondrocytes are often interspersed near chondrocytes, especially toward the periphery of a fibrocartilaginous structure or within transitional zones between dense connective tissue and cartilage. In regions where fibrocartilage merges with tendon or ligament, there is a gradual phenotypic and matrix transition. Fibrocartilage commonly forms at entheses (tendon-bone attachment sites), where it participates in a graded interface between soft and hard tissues [36]. The overall cell density in fibrocartilage is relatively low, although certain regions, such as the outer rim of the meniscus, can exhibit higher cell densities [41]. The structural organization and mixed cellular composition of fibrocartilage equip it to resist both tensile and compressive forces, making it uniquely suited for mechanically demanding interfaces in the musculoskeletal system.

Fibrocartilage typically lacks a perichondrium because it usually contacts other connective tissues directly. For example, the outer region of the knee meniscus blends into the joint capsule (a vascularized fibrous tissue), and the annulus fibrosus of the IVD attaches to the vertebral endplates and surrounding ligaments [6]. These adjacent tissues can provide blood supply at the margins. For example, the outer one-third of the knee meniscus has a sparse capillary penetration from the synovial capsule, and the outer rim of the annulus fibrosus has blood vessels from vertebral attachments. However, the inner one-third of the knee meniscus is avascular, as well as the inner portions of the annulus fibrosus. These regions rely on diffusion for nutrients often aided by mechanical loading cycles [42]. Because there is no perichondrium encapsulating most fibrocartilages, nutrient supply and growth patterns differ from those of hyaline cartilage. Regions near vascularized tissue tend to heal naturally compared to regions far away from the vascularized regions [43].

Many fibrocartilages exhibit a laminar architecture adapted to their specific function. In the annulus fibrosus, there are ~ 15–25 concentric lamellae of collagen, with adjacent lamellae oriented at alternating angles (±30° to the horizontal), forming a plywood-like structure [44]. Fibrochondrocytes reside at the boundaries of these lamellae and help maintain them. In the knee meniscus, collagen is organized predominantly in circumferential fibers (arcing from anterior horn to posterior horn), and under compression, compressive forces are distributed as hoop stresses [3]. Radial fibers also contribute to the distribution of contact pressure and tensile stresses to prevent excessive deformations [45]. Fibrochondrocytes align along the collagen bundles in these directions. At the pubic symphysis, fibrocartilage is arranged in layers sandwiched between thin hyaline cartilage plates on the pubic bone ends [46]. Thus, fibrocartilage often forms a transition or a bonding layer between different tissues.

1.2.3 Functions of Fibrocartilage

The primary functional role of fibrocartilage is to provide tensile strength and toughness in structures that also experience compression [6]. Fibrocartilage's high type I collagen content gives it the ability to resist stretching and tearing. For example, the knee menisci carry significant loads when the knee bears weight. The circumferential collagen fibers resist this by pulling back like a tightened belt ensuring the meniscus does not extrude [41]. This allows the menisci to transmit and distribute loads across the tibiofemoral joint, protecting articular hyaline cartilages. Likewise, the annulus fibrosus of the IVD contains the nucleus pulposus. Under compression, the nucleus pulposus pressurizes, and the annulus's collagen rings prevent it from bulging excessively or herniating [40]. Thus, fibrocartilage converts compressive forces into tensile forces within its collagen network. Because fibrocartilage's proteoglycan content is low, it is not as strong in compression as hyaline cartilage. For example, if a joint surface of hyaline cartilage is replaced by fibrocartilage as sometimes happens in repair, it is less able to handle repeated compression [35]. Clinical observations show that fibrocartilaginous repair tissue in a defect, e.g., after microfracture surgery in a knee, tends to wear out faster under weight-bearing, because it lacks the robust compressive properties of native hyaline cartilage [35]. However, fibrocartilage is indispensable where tensile durability is needed. For example, the glenoid and acetabular labra (rim fibrocartilages of the shoulder and hip) deepen the sockets and stabilize the joints and must resist tearing despite repeated motion and occasional trauma [47, 48]. The fibrocartilage menisci of the knee not only distribute load but also aid in joint congruency and stability. They can sustain complex stresses as the knee flexes and rotates. Fibrocartilage integrates with surrounding tissues to also function as a binding and anchoring structure, creating a gradient from soft to hard tissue that reduces stress concentrations [36]. The collagen fibers of the tendon gradually transition to fibrocartilage with some mineralization before merging with bone. This continuity, enabled by fibrocartilage, is functionally important. It makes the attachment much more robust and less prone to failure than a direct tendon-to-bone junction. The unique composite of fibrocartilage endows fibrous strength and resilience critical for the biomechanical function of many joints. It is biologically adapted for situations where both high pressure and high tension occur together, not ideal for hyaline cartilage or tendon alone. The presence of predominantly type I collagen also allows fibrocartilage to be strongly anchored to bone with collagen fibers from fibrocartilage directly penetrating bone tissue, providing secure fixation in places like the meniscal attachments to the tibia or the annulus fibrosus to the vertebrae. In summary, fibrocartilage functions as a tough composite material, guarding against failure in high-stress junctions and cushioning areas.

1.3 Elastic Cartilage

Elastic cartilage, derived from the ancient Greek elastikós (ἐλαστικός), meaning "springy" or "flexible," is structurally similar to hyaline cartilage but distinguished by the presence of abundant elastin fibers in its matrix [2, 11]. This type of cartilage provides flexible support. It can bend and deform repeatedly and spring back to its original shape. Elastic cartilage is found in specific locations that require shape retention combined with flexibility: the external ear (auricle or pinna) and ear canal, the epiglottis (the flap in the throat that covers the airway during swallowing), the auditory (Eustachian) tube connecting the middle ear to the pharynx, and certain cartilages of the larynx (such as the cuneiform and corniculate cartilages). In these locations, rigid hyaline cartilage would be too brittle, and fibrocartilage would lack the needed shape memory. Elastic cartilage fills the niche by providing a firm framework that is also highly deformable. Fresh elastic cartilage has a "yellowish tinge" because of the elastin content [7]. Under H&E staining, elastic cartilage looks like hyaline cartilage (elastin fibers are not vividly stained by H&E), so special stains like Verhoeff's stain and orcein stain are used to visualize the elastin fibers that surround the territorial matrix and form a band, branching into the interterritorial matrix [49, 50]. The ECM of elastic cartilage containing elastin enables it to maintain structural integrity while accommodating frequent deformation, making it essential in tissues that must bend yet retain their shape.

1.3.1 Biochemical Composition of Elastic Cartilage

The ECM of elastic cartilage is similar to hyaline cartilage in terms of its basic constituents [2]. It has a high content of type II collagen fibrils and aggrecan-rich proteoglycans forming a hydrated gel. Without an elastin-specific stain, its territorial and interterritorial matrix, lacunae, and chondrocytes all resemble those of hyaline cartilage. The critical difference is the addition of a dense network of elastin fibers interwoven with the collagen. The wet weight of elastin in elastic cartilage has been reported to be around 15% [50, 51]. These fibers are composed chiefly of the protein elastin deposited on a scaffold of fibrillin-rich microfibrils. Elastin fibers can stretch and recoil, which imparts elastic cartilage the ability to deform under force and then recover shape when the force is removed [52]. The type II collagen provides baseline tensile strength and shape maintenance, while the elastin fibers endow the tissue with spring-like properties. The proteoglycan content of elastic cartilage is slightly less than that of hyaline cartilage.

Minor collagen types such as type IX collagen and type XI collagen present in hyaline cartilage can be found in elastic cartilage, helping to crosslink fibrils to other matrix components [11]. With elastin stains, a fine to coarse network of elastin fibers that weave through the matrix is revealed. Elastic cartilage also typically has a perichondrium except at free surfaces like the inner side of the epiglottis [53]. The perichondrium around elastic cartilage provides a source of new chondrocytes and also likely contributes to elastin fiber maintenance. It has been reported that elastin is highly stable and undergoes minimal turnover for over 70 years [51]. In summary, the ECM of elastic cartilage contains components of hyaline cartilage and elastin that endow it with support and flexibility.

1.3.2 Organization of Elastic Cartilage

The cellular organization of elastic cartilage is similar to hyaline cartilage, with some differences in cell density. Chondrocytes in elastic cartilage tend to be more abundant and often slightly larger than in hyaline cartilage [2]. Histologically, elastic cartilage has less apparent cell-free matrix space, and the cells appear packed more closely. Chondrocytes reside in large lacunae and are surrounded by elastin fibers in the pericellular area [54]. There is typically no distinct zonal organization in elastic cartilage, unlike articular hyaline cartilage. The arrangement of fibers may vary (e.g., fibers may be slightly more concentrated near the perichondrium), but, overall, the structure is homogeneous. Elastic cartilage is enveloped by a perichondrium in most cases. For example, the auricular cartilage of the ear has a surrounding perichondrium [55]. This perichondrium is crucial during injury. If the ear's perichondrium is stripped as in an auricular hematoma from trauma, the cartilage can be starved of nutrients and develop necrosis leading to deformity like the "cauliflower ear" [56]. Surgeons work to preserve or repair the perichondrium in ear surgeries for this reason. It provides a vascular supply through diffusion and cartilage progenitor cells for healing [7]. The perichondrium of elastic cartilage has a fibrous outer layer and a chondrogenic inner layer like that of hyaline cartilage, capable of appositional growth [30].

1.3.3 Functions of Elastic Cartilage

Elastic cartilage's function is to provide flexible, elastic support. It maintains the shape of structures but permits a high degree of deformation. For example, the auricle of the ear needs to hold its shape to funnel sound, yet it must bend if pressed and then recoil back to its functional shape. Elastic cartilage in the epiglottis keeps the airway open during breathing, flexes to cover the larynx during swallowing, and then returns to its shape [57]. The mechanical behavior of elastic cartilage is dominated by the elastin fiber network. In bending or compression, the elastin fibers stretch and store energy, then release it to spring back once the force is gone. The presence of elastin gives this tissue a rubber-like resilience. Meanwhile, the type II collagen and proteoglycans ensure that elastic cartilage still has stiffness. The tensile strength of elastic cartilage is moderate, similar to that of hyaline cartilage and weaker than fibrocartilage. Another feature of elastic cartilage is that it tends not to calcify with age [30]. Hyaline cartilage like costal cartilages or the laryngeal cartilages often undergoes calcification or ossification in older adults, losing flexibility. Importantly, ears and nose tips remain pliable even in old age. Elastic cartilage can serve as a protective yet pliant framework.

Elastic cartilage has a regenerative potential associated with progenitor cells in the perichondrium, especially at the periphery where vascular supply is present [30]. A torn elastic cartilage (like a partial ear injury) can heal if the perichondrium remains intact because new chondroblasts from the perichondrium can fill in the defect by appositional growth. Nevertheless, large defects or complete avulsions of elastic cartilage still heal poorly, as cartilage in general has limited repair capacity. The function of the perichondrium is significant, and surgeons repairing the ear take care to suture the perichondrium. An auricular hematoma, which separates the perichondrium from cartilage, can lead to cartilage death and deformity if not promptly drained. In summary, elastic cartilage functions to provide shape with suppleness. It gives form to the external ear and epiglottis in a way that is resilient to frequent bending. Its elastin-collagen-proteoglycan composite matrix allows it to repeatedly store and release elastic energy (like a spring), a property vital for structures that move but must return to a set shape. This makes elastic cartilage ideal for roles that demand structural integrity under deformation. Unlike hyaline cartilage in joints, elastic cartilage is not primarily a low-friction bearing surface, nor is it a major weight-bearer; instead, its hallmark is reversible deformability.

1.4 Lipocartilage

Lipocartilage is a recently characterized type of cartilage containing fat-laden cells called lipochondrocytes [9]. Previous studies had observed in histology the accumulation of intracellular lipid droplets in the mouse and rat external ear cartilages without further characterization [58–61]. In the recent study, lipid-containing chondrocytes were observed in multiple mammalian species, including marsupials, rodents, and bats [9]. In humans, lipid-containing chondrocytes were observed in fetal nasal, auricular, and thyroid and epiglottic cartilages by 20 weeks' gestation [9]. Additionally, there have been reports of lipid-containing chondrocytes in adult human auricular cartilage [50]. Because of the novelty of lipocartilage, there remains much work in characterizing and understanding lipocartilage and lipochondrocytes in humans as well as other species.

1.4.1 Biochemical Composition of Lipocartilage

Although limited information has been published on the biochemical composition of lipocartilage, it has been shown to have distinct compositional differences from adipose tissue. In a comparison between 1-month-old mouse ear lipocartilage and inguinal white adipose tissue, lipocartilage contained approximately half the triglyceride content and three times the amount of fatty acid esters of hydroxy fatty acids [9]. This suggests that lipid vacuoles in lipocartilage arise through distinct metabolic pathways. The ECM of adipose tissue was composed predominantly of type I and III collagens (97%), whereas lipocartilage contained only 62.2% of these collagens and instead included additional ECM proteins, such as type II, VI, and VIII collagens, elastin, and myocilin [9]. Vimentin, an intermediate filament protein known to form cage-like structures around early lipid vacuoles, was also found to be 15 times more abundant in lipocartilage than in adipose tissue. Further studies are needed to better define how lipocartilage differs from other cartilage types and to characterize developmental changes in its composition.

1.4.2 Organization of Lipocartilage

Lipocartilage organization varies across species, and further studies are needed to elucidate both its structural organization and the developmental changes that occur over time. In rodents, lipocartilage is characterized by lipochondrocytes containing large lipid droplets, and these cells are observed as distributed throughout the tissue. In human fetal cartilage, lipochondrocytes with prominent lipid vacuoles have been identified in the ear, nasal, and thyroid and epiglottic cartilages at 20 to 21 weeks of gestation, although they do not comprise the entire tissue [9]. In certain bat species, lipocartilage exhibits a more complex architecture. Lipochondrocytes are sparsely distributed at one end and form periodic ridges containing stacked lipochondrocytes, producing a ruffled ear morphology thought to aid in sound perception. More detailed analyses are necessary to uncover the functional significance of lipocartilage in both development and mature tissue physiology.

1.4.3 Functions of Lipocartilage

The functions of lipocartilage will be better understood with further investigation into its biochemical and mechanical properties, particularly given the diversity observed across species and the transient nature of lipocartilage in humans. In the recent study, lipocartilage exhibited unexpectedly robust mechanical properties despite being composed of large, lipid-filled cells and containing minimal ECM [9]. A deeper understanding of its heterogeneous structure, including the role of lipid vacuoles in providing compliance, structural support, and shape retention, is essential to elucidate how lipocartilage contributes to tissue function during development and in mature states.

1.5 Cartilage Development

Cartilage development is a complex and finely regulated biological process. Much of our foundational understanding has been drawn from classical embryological models, especially studies of limb bud and neural crest formation in the chick embryo, frequently employing the quail-chick chimera system [62–65]. Subsequent advances have emerged from mouse models, where genetic tools such as Cre/loxP-based fate mapping and conditional knockout have enabled detailed lineage tracing and mechanistic studies [66–68]. Additional insights into both development and regeneration have come from non-mammalian vertebrate systems like zebrafish, Xenopus, and anole lizards, which offer complementary advantages in imaging, gene editing, and evolutionary context [69–72]. Fundamental processes, including mesenchymal condensation, chondrogenesis, and matrix deposition, are conserved across cartilage types. Accordingly, this section presents cartilage development in chronological order, beginning with embryogenesis and continuing through postnatal maturation toward skeletal maturity. The mechanical and physicochemical influences on cartilage development will be discussed briefly, followed by a discussion of aging, particularly the structural and functional decline that occurs during late adulthood.

1.5.1 Embryogenesis and Cartilage Maturation

Cartilages undergo a complex developmental process across the human lifespan (Fig. 1.4). They first arise during embryogenesis as immature structures and undergo substantial growth and remodeling throughout infancy, childhood, and adolescence [2]. After skeletal maturity is reached in early adulthood, cartilage properties tend to remain relatively stable through early to mid-adulthood (approximately ages 20 to 45). However, beginning in late adulthood (around age 50 and beyond), cartilages exhibit progressive decline in function and become increasingly susceptible to degeneration. An understanding of this developmental process is essential for developing tissue engineering strategies aimed at recreating biomimetic (native-like) cartilages and optimizing treatment responses to tissue engineering therapies.

Embryonic origins for the different types of cartilages throughout the body differ. In vertebrate development, cartilages arise from two germ layers: the mesoderm and, indirectly, the ectoderm through the neural crest. The mesoderm, located between the ectoderm and endoderm, differentiates into paraxial, intermediate, and lateral plate mesoderm

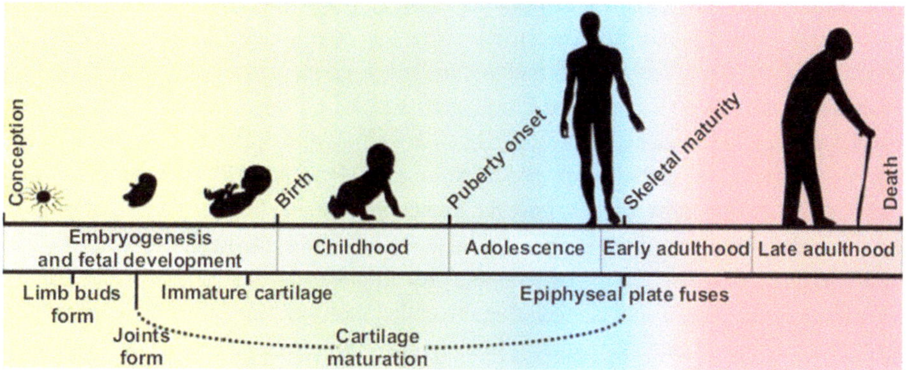

Fig. 1.4 Cartilage development stages across the human lifespan. Cartilage originates during embryogenesis and gives rise to an early, immature tissue. Through infancy, childhood, and adolescence, the tissue undergoes significant growth and maturation, characterized by structural and biochemical remodeling until reaching skeletal maturity in early adulthood. During early to mid-adulthood (approximately ages 20 to 45), cartilage properties remain relatively stable. In late adulthood (around age 50 and beyond), cartilage and chondrocyte function may gradually decline, often in association with age-related degeneration or disease. Diagram not to scale. Reproduced with permission [2]

[73, 74]. The paraxial mesoderm, through its sclerotome, gives rise to axial cartilages such as the costal cartilages, annulus fibrosus of the IVD, knee menisci, and pubic symphysis [75]. The lateral plate mesoderm, particularly its somatic layer, contributes to appendicular and visceral cartilages, including the articular cartilage of synovial joints (e.g., the knee, hip, and shoulder), tracheal and bronchial rings, and labrum. In contrast, craniofacial cartilages originate from neural crest cells, which emerge at the border of the neural plate and undergo epithelial-to-mesenchymal transition before migrating into the pharyngeal arches [73]. These neural crest-derived cartilages include the nasal cartilage, auricular cartilage, Meckel's cartilage, and cartilages of the TMJ [76]. Specifically, both the TMJ disc and the mandibular condyle develop from neural crest-derived mesenchyme [77]. Thus, while most limb and trunk cartilages originate from the mesoderm, many craniofacial cartilages arise from the neural crest. Understanding these developmental pathways is important for establishing tissue engineering strategies that aim to recapitulate these native cartilages.

Among the major developmental pathways for cartilage formation, the limb bud, which arises from the lateral plate mesoderm, has been one of the most extensively studied models. In humans, limb development begins around week 4.5 of gestation, with cartilage differentiation occurring by weeks 7 to 9 [2]. The limb bud mesenchyme, derived from lateral plate mesoderm and encased in an ectodermal cap, contains early progenitor cells that undergo mesenchymal condensation at sites where skeletal elements will form [78]. This condensation is mediated by adhesion molecules such as neural-cadherin (N-cadherin) and neural cell adhesion molecule (N-CAM), and regulated by transforming growth factor-beta

(TGF-β) and bone morphogenetic proteins (BMPs) [79]. The condensed mesenchymal cells initiate chondrogenic differentiation under the direction of SRY-box transcription factor 9 (SOX9), along with SOX5 and SOX6, which cooperatively activate key cartilage matrix genes such as *COL2A1* (type II collagen) and *ACAN* (aggrecan) [67]. These events produce the cartilage anlage, a primitive hyaline cartilage template that expands through matrix deposition and chondrocyte proliferation [80]. Peripheral mesenchymal cells surrounding the anlage differentiate into the perichondrium, a fibrous layer that supports appositional growth and provides progenitor cells [81]. Mesenchymal condensation and chondrogenesis establish the initial templates for limb cartilages.

At future joint sites, a specialized structure called the interzone forms between adjacent cartilage anlagen and plays a pivotal role in joint formation (Fig. 1.5). This region arises from condensed mesenchymal cells and differentiates into a dense, layered population that establishes the nascent joint boundary [82]. Interzone cells secrete signals, such as wingless-type MMTV integration site family member 9A (Wnt9a), growth differentiation factor 5 (GDF5), and Noggin, to locally inhibit chondrogenesis and permit joint cavitation [83, 84]. These cells undergo selective apoptosis, leading to the formation of small cavities within the interzone that coalesce to form the synovial joint cavity [80]. The interzone also contributes cells to joint-associated structures including the synovium, joint capsule, and knee menisci, while chondrocytes at the periphery of the interzone give rise to the permanent articular cartilage that caps long bones [83, 84]. These articular chondrocytes do not undergo hypertrophy or ossification, preserving their long-term function as hyaline cartilage that covers the joint surfaces.

In contrast, in skeletal regions destined for ossification (bone formation), chondrocytes in the center of the anlage exit the cell cycle and undergo hypertrophic differentiation (Fig. 1.6). This process is marked by enlargement and secretion of type X collagen, alkaline phosphatase, and vascular endothelial growth factor (VEGF), which initiate matrix mineralization and vascular invasion [85]. This triggers endochondral ossification, beginning in the primary ossification center of the diaphysis during fetal development [86]. Later, secondary ossification centers appear in the epiphyses, preserving a cap of permanent hyaline cartilage at the joint surface and forming the epiphyseal growth plate, a transient cartilage zone responsible for postnatal bone elongation. While much of the embryonic cartilage is ultimately replaced by bone, specific regions, including tracheal rings, auricular cartilage, and articular surfaces, retain their hyaline cartilage identity. Notably, cells at the surface of developing articular cartilage flatten and begin expressing lubricin (PRG4), forming the superficial zone essential for joint lubrication and shear resistance. By the end of embryogenesis and early postnatal life, the structural blueprint of each joint, comprising a joint cavity, synovial lining, permanent articular cartilage, and supporting connective tissues, is largely established.

Following birth, cartilages undergo significant structural and biochemical remodeling, continuing until skeletal maturity is reached in early adulthood. During infancy and childhood, cartilage tissues grow and adapt in response to increasing mechanical demands,

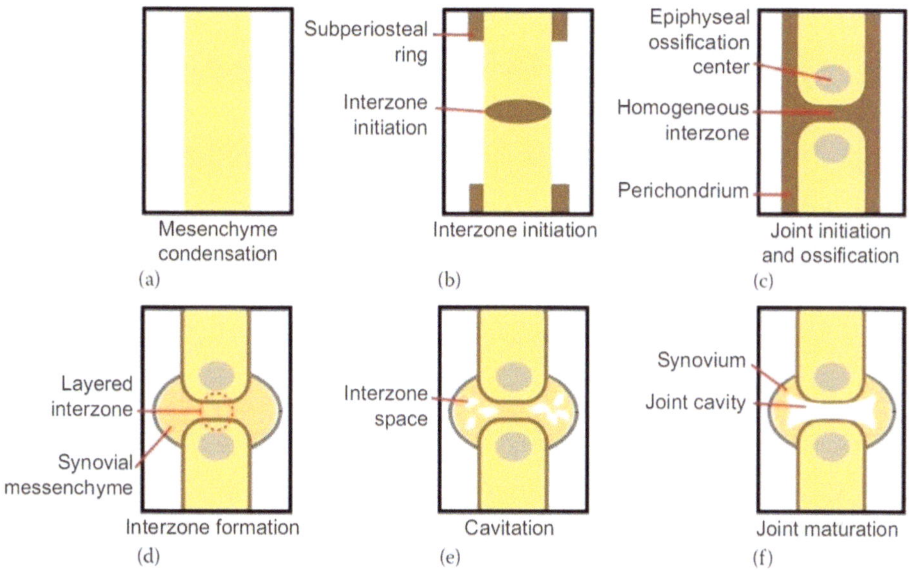

Fig. 1.5 Process of joint development. Starting from mesenchymal condensation **a**, the joint interzone begins to form and delineates the future joint space **b**. As ossification proceeds in the epiphyseal centers **c**, the interzone differentiates into a layered structure, and synovial mesenchyme emerges **d**. Within the interzone, selective apoptosis creates distinct cavities that coalesce to form the nascent joint cavity **e**. This process establishes the initial structural framework of the joint **f**, which will continue to mature and remodel until skeletal maturity is achieved. Reproduced with permission [2]

nutritional cues, and hormonal changes [87, 88]. As the musculoskeletal system begins to support body weight and locomotion, joint usage promotes the development of depth-dependent zonal architecture, particularly in articular cartilage, where collagen fibrils align in response to directional stress [89]. In the superficial zone, chondrocytes express lubricin, and collagen fibers orient parallel to the surface, reducing friction and enhancing shear resistance [89–91]. These changes also reduce tissue permeability and increase both tensile stiffness and dynamic modulus, progressively shifting load-bearing roles to deeper matrix zones as the tissue matures [91]. In deeper zones, mechanical compression stimulates proteoglycan synthesis, further enhancing compressive resistance.

Chondrocyte density is high during early life, and the cells remain metabolically active and mitotically competent throughout the postnatal growth phase [2]. This is particularly evident in the epiphyseal growth plates, where proliferating chondrocytes contribute to longitudinal bone growth [92]. Non-articular cartilages such as the trachea and costal cartilages also expand with overall body size and decrease in chondrocyte density [81]. As adolescence progresses and skeletal maturity is reached, the growth plates ossify, chondrocyte proliferation ceases, and the tissue enters a phase of greater structural stability. By the

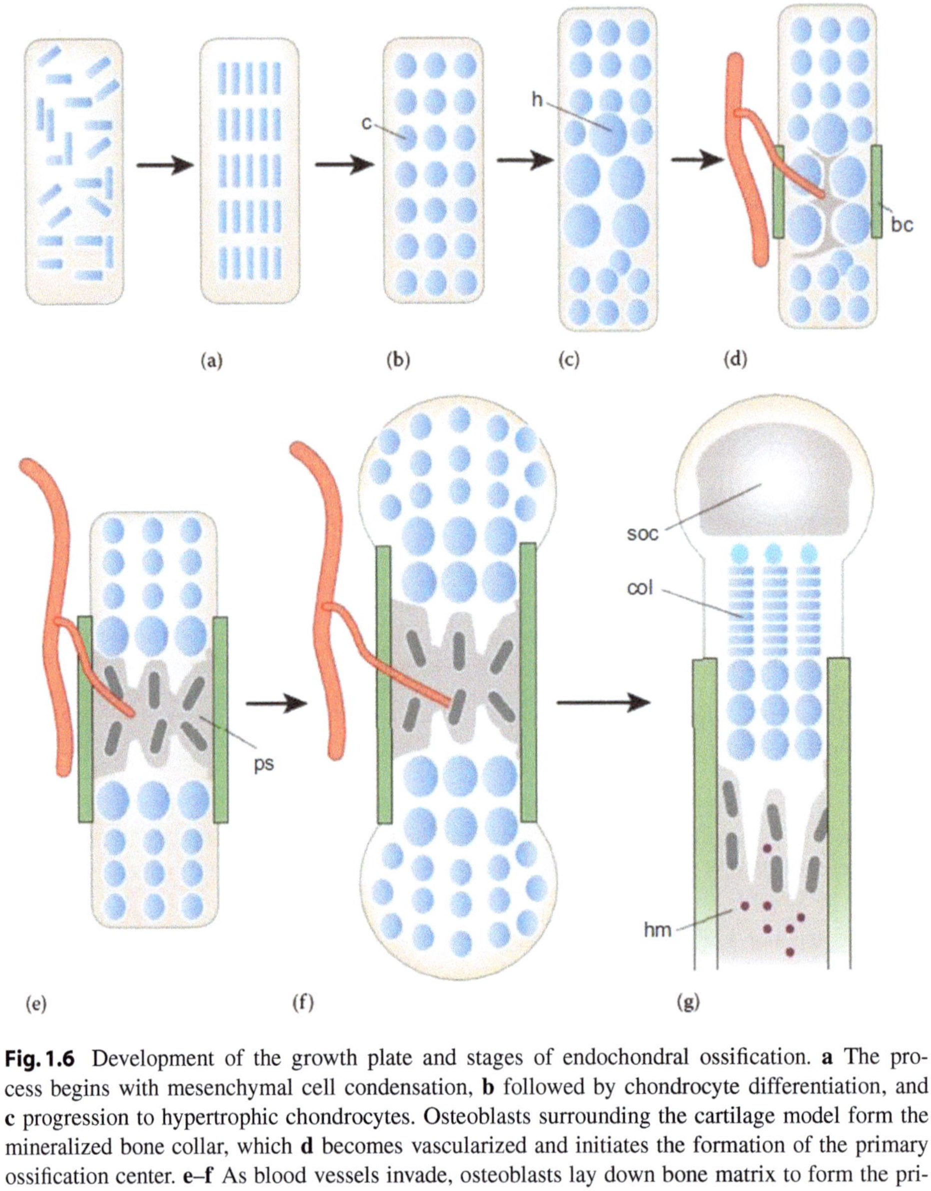

Fig. 1.6 Development of the growth plate and stages of endochondral ossification. **a** The process begins with mesenchymal cell condensation, **b** followed by chondrocyte differentiation, and **c** progression to hypertrophic chondrocytes. Osteoblasts surrounding the cartilage model form the mineralized bone collar, which **d** becomes vascularized and initiates the formation of the primary ossification center. **e–f** As blood vessels invade, osteoblasts lay down bone matrix to form the primary spongiosa, the precursor of trabecular bone. **g** Continued longitudinal bone growth occurs through proliferating columnar chondrocytes, while secondary ossification centers appear at the epiphyses. With skeletal maturation, hematopoietic marrow occupies the central medullary region. Abbreviations: c = chondrocyte, h = hypertrophy, bc = bone collar, ps = primary spongiosa, col = columnar chondrocytes, hm = hematopoietic marrow, soc = secondary ossification centers. Adapted [93] and reproduced [2] with permission

end of puberty, cartilage in most regions exhibits the compositional and organizational features of adult tissue. Collagen and proteoglycan content stabilize, matrix turnover slows, and zonal architecture becomes fully defined. From this point through midlife (approximately ages 20 to 45), cartilage remains in a relatively steady-state condition, with stable mechanical properties and limited remodeling.

In summary, cartilage development from embryogenesis through skeletal maturity is characterized by tightly orchestrated transitions, mesenchymal condensation, chondrogenesis, ossification, joint specialization, and maturation, all of which contribute to the structure and function of cartilages throughout the body. Most maturation is complete by adolescence, and cartilages then remain largely unchanged until the onset of aging, or the period following skeletal maturity [81].

1.5.2 Mechanical and Physicochemical Regulation

Cartilage development is profoundly influenced by biomechanical and biochemical cues in the embryonic and postnatal environment. One critical factor is biomechanical stimulation. Even before birth, the movements of the fetus play an essential role in normal skeletal and joint development [94]. Embryonic joints require intermittent mechanical loading to develop properly. Experiments in which fetal movement is reduced or eliminated (for example, neuromuscular paralysis in animal models) result in joint malformations or even joint fusions due to inappropriate ossification of what would have formed cartilages [95, 96]. In normal development, muscle contractions and movements of the limbs generate dynamic mechanical forces on developing cartilage anlagen and joints. These forces promote the formation of the synovial joint cavity and help direct cells at the joint surface to form articular hyaline cartilage rather than bone. They also influence the growth and shape of developing bones and cartilages. For example, mechanical stimuli upregulate synthesis of cartilage matrix and can orient the collagen fiber architecture [97, 98]. After birth, as the infant begins to bear weight and use joints, physiological loading continues to shape the cartilages. The collagen network of articular cartilage aligns according to stress with a characteristic parallel-to-surface orientation in the superficial zone due to shear forces from joint motion. Chondrocytes increase production of proteoglycans in response to compressive loads, contributing to tissue thickening and maturation. Conversely, abnormal or excessive mechanical stress can impair development. For example, uneven loading can lead to altered cartilage thickness or organization, while lack of mechanical stimulation (such as prolonged immobilization) can cause thinning of cartilages [99, 100]. Overall, an appropriate mechanical environment is a requisite for developing functional, biomechanically robust cartilage structures, a principle that has been carried into tissue engineering strategies as further discussed in Chap. 3.

In parallel with biomechanical cues, the oxygen tension and vascular environment are pivotal in cartilage development. Cartilages are avascular tissues in their mature state, and, even during development, the interior of the anlagen is relatively hypoxic (low in oxygen content) [101]. Multiple studies have demonstrated that hypoxic conditions (2–5% O_2, compared to ~ 20% O_2 in air) encourage mesenchymal stromal/stem cells (MSCs) to differentiate into chondrocytes and to maintain cartilage matrix production [101–103]. This is consistent in the body, where the center of a growing cartilage template can exist at oxygen levels as low as 1–2%. The low-oxygen environment helps cartilage by suppressing unwanted hypertrophy and calcification until the proper time. Low oxygen tension stabilizes hypoxia-inducible factor 1α (HIF-1α) in cartilage-derived progenitor cells (CPCs) and developing chondrocytes, which, in turn, promotes the expression of cartilage genes and inhibits vascular invasion. Only when hypertrophic chondrocytes begin to secrete pro-angiogenic factors such as VEGF and the surrounding matrix is mineralized will blood vessels penetrate the cartilage. If cartilage becomes too oxygenated (for example, if a region becomes vascularized prematurely), ossification or degenerative changes may be triggered. Thus, the avascular nature and resulting hypoxia of cartilage are key to its phenotypic stability. In development, the joint interzone remains avascular, and those cells resist chondrogenic hypertrophy, thereby preserving the articular cartilage [104]. From a tissue engineering perspective, hypoxia or low oxygen culture conditions are often used to enhance cartilage formation in vitro, mimicking the developmental milieu.

In summary, both biomechanical forces and physicochemical conditions, such as oxygen availability, shape cartilage development. These factors ensure that cartilages form in the right locations (e.g., at joint surfaces) and with proper structure. The interplay of biomechanical loading and low oxygen tension during development results in well-organized, resilient cartilage tissues. Deviations from these conditions can lead to developmental abnormalities or suboptimal cartilage properties. The understanding of these environmental influences has informed tissue engineering strategies, as further discussed in Chap. 3.

1.5.3 Aging

Cartilages undergo noticeable changes during aging, as defined here as the period after skeletal maturity. Aging affects the structure, composition, cellular activity, and mechanical properties of the mature cartilages. Although these changes vary across anatomical sites and cartilage types, they share several common features: reduced cellularity, diminished metabolic activity, matrix stiffening, and a decline in repair capacity. These age-related considerations are especially important in tissue engineering, where functional demands differ by age.

Aging is accompanied by a marked reduction in chondrocyte density and proliferation, along with increased rates of apoptosis. Chondrocytes in aged tissue exhibit a diminished response to anabolic stimuli such as insulin-like growth factor 1 (IGF-1), often because of elevated levels of growth factor-binding proteins that limit signaling [105]. These cellular changes make aged cartilage less resilient to both injury and mechanical overload and more vulnerable to degenerative processes [106]. Telomerase activity is minimal in adult cartilage, and chondrocytes undergo a dramatic decline in mitotic activity. By some estimates, telomerase activity is reduced by ten-fold over the course of 80 years, reflecting increasing cellular senescence over the human lifespan [107].

Matrix composition changes in aging. Collagen fibers accumulate non-enzymatic cross-links with age, particularly advanced glycation end-products (AGEs), which stiffen the network and reduce its toughness [106, 108]. They make cartilages more brittle and susceptible to cracking under shear or impact. Collagen alignment becomes disorganized, and birefringence, a marker of fiber orientation, diminishes [109]. Regulatory collagens such as type XI decline in abundance after adolescence, indicating a shift in collagen fibril regulation during maturation. These changes are especially pronounced in load-bearing regions, where microdamage accumulates over decades. In many individuals, localized matrix degradation or calcification develops even in the absence of diagnosed disease. Proteoglycans also undergo substantial age-related changes. The total content of aggrecan declines, and aggrecan undergoes post-translational modifications that alter the pattern of glycosylation and sulphation of GAG chains [110]. Shifts in sulfation patterns lead to an increased ratio of chondroitin-6-sulfate to chondroitin-4-sulfate and higher proportions of keratan sulfate [111]. Link protein, which stabilizes aggrecan-hyaluronan aggregates, fragments over time, further weakening stability [112]. As a result, aged cartilage holds less water and is less able to resist compressive loads.

Different cartilages exhibit distinct aging trajectories. In articular cartilage, thinning, matrix loss, and surface fibrillation are common with age, resulting in reduced ability of tissues to recover from minor injuries [113]. Calcification may occur in focal areas, and the zonal organization becomes less distinct. In costal cartilage, calcification and even ossification are common in the elderly, leading to decreased thoracic flexibility [114]. The knee meniscus and annulus fibrosis of the IVD lose hydration and show increased fragmentation of collagen and proteoglycan components, contributing to reduced resilience to compression and higher risk of degeneration or herniation [44]. Cartilages of the TMJ, including the disc, may also degenerate over time, exhibiting thinning, fragmentation, or displacement [115]. Elastic cartilages, such as those of the auricle and epiglottis, experience subtle degradation with age, including fragmentation of elastin fibers and loss of elasticity, although they often retain structural integrity better than load-bearing cartilages [50].

Systemic factors further influence cartilage aging. Hormonal changes, particularly estrogen deficiency after menopause, have been shown to impair cartilage mechanical properties [116]. Diabetes accelerates matrix glycation and increases oxidative stress, both of which negatively affect cartilage integrity [117]. Pharmacologic treatments, such as repeated intra-articular corticosteroid injections, can degrade cartilage over time [118]. Although not all aged cartilage is clinically diseased, the cumulative effects of aging render the tissue less responsive to repair and more susceptible to degeneration.

In summary, aging after skeletal maturity transforms cartilage from a dynamic, adaptive tissue into one that is structurally stiffer, biologically quiescent, and mechanically less resilient and fragile. For tissue engineering, this means that successful strategies must either replicate the properties of youthful cartilage or counteract the deficiencies that accumulate with age in order to restore long-term function. These progressive changes in cartilages, especially in mechanically loaded regions like the knee, spine, and TMJ, contribute to cartilage injury and degeneration, which are explored further in the next section.

1.6 Cartilage Injury and Degeneration

Cartilage injury and degeneration are major contributors to musculoskeletal disease, imposing substantial healthcare costs and negatively impacting quality of life and healthy years lived. The intrinsic healing capacity of cartilage is limited (e.g., because of lack of vascularity), making damage difficult to arrest or reverse. The response to injury and degeneration varies depending on cartilage type and anatomical location, necessitating targeted clinical and tissue engineering approaches. This section discusses the injury and degeneration of hyaline cartilage, fibrocartilage, and elastic cartilage, with attention to prevalence and anatomical specificity. Injury and degeneration of lipocartilage are not discussed in this section, as it represents a recently identified cartilage subtype and remains to be characterized. Further studies are needed to understand its relation to human pathologies.

1.6.1 Hyaline Cartilage

Damage and degeneration of hyaline cartilage, particularly in synovial joints, represent a major clinical challenge because of their prevalence and cartilage's limited capacity for repair. Articular cartilage defects are identified in roughly 60% of individuals undergoing knee arthroscopy, with incidence rising progressively with age [119]. These defects can range from early surface irregularities, such as fibrillation or microdamage, to more extensive full-thickness defects that reach the bone [3]. Although microdamage may not immediately produce visible matrix disruption, it results in further damage to the collagen network, which contributes to early GAG loss [120]. As the matrix becomes more disorganized, water content increases, altering the tissue's mechanical properties [121]. Changes in load distribution due to microdamage can concentrate mechanical stress in localized areas, potentially extending matrix damage and transferring a disproportionate load to the underlying bone. This cascade may contribute to thickening of both the subchondral bone and the calcified cartilage layer, especially as the surface cartilage thins [122]. Because cartilage lacks innervation, these microinjuries may not induce pain, allowing degenerative processes to worsen unnoticed by patients [123]. Although initial GAG loss may trigger increased activity in resident chondrocytes, this response is generally insufficient to restore matrix integrity, resulting in a continued decline in GAG content, increased surface wear, and eventual fissure formation.

Partial-thickness defects, those that involve just the cartilage layer, can proceed from cartilage microdamage or from trauma, improper loading, or foreign bodies. Without blood, the intrinsic metabolic activity after such an injury is insufficient to result in adequate repair, leading to the eventual development of full-thickness and osteochondral defects that involve the subchondral bone [123]. Osteochondral defects are lesions that cross the tidemark and penetrate the underlying bone. Though growth factors and progenitor cells are recruited from the bone's vasculature, there is impaired functionality of the repair tissue, which consists of a mix of fibro- and hyaline cartilages, resulting in eventual degeneration into osteoarthritis (OA), a chronic degenerative joint disease [35]. Over time, untreated focal or localized defects can progress to further degeneration and ultimately to OA. In the United States, OA affected over 32.5 million adults, accounting for 9.8% of the country's population in 2020 [124]. The annual direct medical costs of OA were estimated at $72 billion between 2008 and 2011 in the US [125]. As the 14th leading cause of years lived with disability worldwide, OA is a major target for tissue engineering [126].

Other hyaline cartilage tissues also experience degeneration. Costal cartilage calcification is prevalent in approximately 26 to 50% of elderly individuals, reducing thoracic compliance and contributing to respiratory restriction [114]. Nasal septal cartilage is susceptible to trauma and surgical resection. Once damaged, it exhibits poor intrinsic healing and often requires graft-based reconstruction. For example, an imaging study found 2.1% of adults over age 21 had nasal septal perforations [127]. The nose is the most frequent site of facial skin cancer, accounting for approximately 36% of cases, and tumor removal often results in damage to nasal cartilage [4]. Additionally, nasal burns are reported in up to 70% of patients treated at civilian burn centers, frequently leading to infection, scarring, and airway compromise due to deformity and contracture. Nasal cartilage injuries are also common among military personnel, particularly from blast and burn-related mid-facial trauma, which affects 44% of combat zone evacuations between 2001 and 2011 [128]. If nasal function is not properly restored, patients may suffer from chronic sinus issues, nosebleeds, headaches, sleep disturbances, and even cognitive impairment [129]. Moreover, nasal deformities can affect self-image, social interactions, and perceived trustworthiness or employability. Reconstructive procedures like rhinoplasty and septoplasty are essential to restore both structure and function. Although nasal cartilage appears to bear less mechanical load than articular cartilage, its deterioration can still disrupt breathing and airway patency.

1.6.2 Fibrocartilage

Fibrocartilages are present in high-load, load-distributing regions of the body and are commonly affected by degenerative disease and trauma. Fibrocartilages include the knee menisci, IVDs, pubic symphysis, and TMJ discs. They have a dense matrix rich in type I collagen and are generally avascular and without perichondrium, properties that severely limit their capacity for repair.

Meniscal injuries are highly prevalent, affecting ~35% of the population above age 50 [130]. They often occur through acute rotation or compressive trauma but can also arise from chronic degeneration [3]. Tears in the outer region of the knee meniscus, the vascularized zone, may exhibit limited healing, but those affecting the inner region rarely repair without intervention [131]. Meniscal degeneration often accompanies early OA and is marked by matrix fraying, proteoglycan loss, and collagen disorganization. These changes diminish load transmission and accelerate articular cartilage wear.

IVD degeneration is nearly universal in older adults. Radiographic evidence of disc space narrowing, annular fissuring, and nucleus pulposus dehydration appears in 57% of individuals over 60 years old. These changes contribute to low back pain, which affects more than 619 million people globally and is the top cause of years lived with disability [132]. Degeneration impairs spinal shock absorption, leads to disc height loss and segmental instability, and may result in disc herniation and nerve impingement.

Temporomandibular disorders (TMDs), which encompass a range of conditions affecting the TMJ, impact an estimated 5–25% of the population [133]. One of the most common manifestations is TMJ disc displacement, reported in up to 70% of TMD cases, with a predominant pattern of anterior displacement [133]. Disc displacement may progress to structural degradation, including thinning or full-thickness perforation, observed in approximately 10–15% of affected individuals. Notably, a study examining the relationship between displacement and defect location found that 53.7% of anteriorly displaced discs developed defects in the lateral third of the disc [134]. Despite the prevalence and clinical importance of these pathologies, therapeutic options remain limited, emphasizing the need for regenerative strategies [115]. The TMJ disc is particularly vulnerable to mechanical damage and degeneration, which can result in symptoms such as pain, joint sounds, and reduced mandibular mobility. In more severe TMD cases, degeneration may extend to include remodeling of the condylar cartilage and subchondral bone, further compromising joint integrity and function.

The pubic symphysis, a fibrocartilaginous joint, can degenerate or become inflamed, particularly in pregnancy or after trauma [135]. Though less commonly a source of chronic pain, symphyseal degeneration can contribute to pelvic instability and is occasionally implicated in persistent postpartum pelvic pain syndromes [136].

1.6.3 Elastic Cartilage

Elastic cartilage is less frequently injured than hyaline cartilage or fibrocartilage but remains clinically relevant. It is found in the auricle, epiglottis, and Eustachian tube, where it provides both flexibility and structural support. Though relatively protected from compressive forces, elastic cartilage can undergo degeneration, especially in exposed locations.

Auricular cartilage is vulnerable to blunt trauma, particularly in contact sports such as wrestling, boxing, and judo [137]. A perichondrial hematoma may form, and if not drained promptly, the underlying cartilage becomes ischemic and fibrotic, resulting in the permanent deformity known as "cauliflower ear." While prevalence data are not well-established, the condition is common among athletes in high-impact sports. For example, ear deformities were found in 55.5% of top athletes in judo [137]. Age-related fragmentation of elastin fibers occurs in all elastic cartilage. In the auricle, this contributes to gradual loss of firmness and shape with aging. In the epiglottis, degradation and occasional calcification may reduce the structure's ability to protect the airway during swallowing, potentially contributing to dysphagia in elderly individuals [138]. Although less studied, degeneration of the Eustachian tube cartilage can impair its function in middle ear pressure regulation; it is suspected to contribute to chronic otitis media in infants and children [139]. Repair of elastic cartilage is particularly challenging due to the need to restore both structural shape and flexibility. No current clinical or tissue engineering strategy fully replicates the biomechanical behavior of native elastic cartilage, making injuries or deficiencies in these regions difficult to treat.

To summarize, cartilage injury and degeneration are prevalent and clinically significant, varying by cartilage type and anatomical location. Hyaline cartilage injuries, especially in joints, are among the most common and are directly implicated in OA, which affects tens of millions of adults. Fibrocartilage degeneration contributes to back pain, knee pain, and TMDs, each with substantial health and economic burdens. Elastic cartilage degeneration is less prevalent but may compromise structural and functional roles in the airway and auditory system. In all types of cartilage, degeneration follows a shared trajectory of chondrocyte apoptosis, matrix breakdown, and mechanical failure. Repair is limited by intrinsic biology, and current clinical strategies often yield suboptimal tissue. These challenges highlight the need for site- and tissue-specific repair approaches in cartilage tissue engineering that account for differences in mechanical function, age-related decline, and disease.

Chapter concepts:

- Cartilages, distributed throughout the body, are specialized connective tissues composed of chondrocytes and ECM rich in collagens and proteoglycans.

- The three main types of cartilage, hyaline cartilage, fibrocartilage, and elastic cartilage, each possess distinct biochemical, structural, and functional characteristics, with lipocartilage recently identified as a fourth type.
- Hyaline cartilage, primarily associated with type II collagen, is the most prevalent cartilage type and forms the articular surfaces of synovial joints, costal cartilages, nasal septum, and respiratory tract.
- Fibrocartilage contains dense type I collagen bundles and is found in IVDs, knee menisci, pubic symphysis, TMJ discs, and entheses, where it resists both compression and tensile forces.
- Elastic cartilage is rich in elastin fibers, allowing for flexible support in the auricle, epiglottis, and Eustachian tube.
- Lipocartilage, a fat-containing cartilage containing lipochondrocytes, provides deformability in soft-tissue regions such as the nose tip and earlobe in some mammalian species.
- Cartilages arise from mesenchymal progenitors originating from paraxial mesoderm, lateral plate mesoderm, and neural crest, depending on anatomical location.
- Chondrogenesis begins with mesenchymal condensation, followed by SOX9-driven differentiation into chondrocytes and deposition of cartilage-specific ECM.
- Articular cartilage does not undergo endochondral ossification and persists as permanent hyaline cartilage in adult joints.
- With age, cartilage becomes thinner, less cellular, and less responsive to growth factors, with decreased proteoglycan content and increased collagen crosslinking.
- Hyaline cartilage degeneration commonly leads to OA, affecting over 32.5 million adults in the United States.
- Fibrocartilage injuries and degeneration contribute to widespread conditions such as low back pain from IVD degeneration, meniscal tears, and TMDs.
- Elastic cartilage is less prone to mechanical injury but shows age-related elastin fragmentation, as seen in epiglottic calcification.
- All cartilage types share limited repair capacity, and current tissue engineering strategies must replicate their complex structure and functions that can withstand diverse biochemical and biomechanical environments to restore function.

2 Anatomy, Structure, and Function of Select Native Cartilages

Rachel C. Nordberg, Takumi Takahashi, Jerry C. Hu, and Kyriacos A. Athanasiou

The three broad categories of cartilages are hyaline cartilage, fibrocartilage, and elastic cartilage, and within each category are numerous distinct cartilaginous structures throughout the body. There are hundreds of joints in the human body, and most of them contain cartilage. Moreover, cartilage is present in other structures such as growth plates, ribs, facial structures, and tracheal structures. Even within a single joint (e.g., knee) or anatomical structure (e.g., nose), multiple cartilage types can exist. Thus, a detailed review of the structure and function of all cartilages within the body is outside the scope of this text. Instead, here we describe select cartilages that have either been studied extensively within the literature or are emerging as important areas of research. These include examples of hyaline cartilage, fibrocartilage, and elastic cartilage. Additionally, we provide an overview of the newly characterized lipocartilage. By providing these representative examples, it is our aim to provide the reader with a basic understanding of the major cartilages within the body.

When contemplating a tissue engineering approach to cartilage repair, it is important to characterize the tissue that will be replaced in terms of its functional characteristics (e.g., biochemical content, biomechanical properties, or quantitative measurements of how the tissue functions). Human characterization data are the most relevant toward designing a clinical tissue-engineered product, but it can be difficult to obtain healthy samples for characterization. As seen below, there is an incomplete understanding of the biochemical and biomechanical characteristics of many human cartilages. Characterization of tissues from relevant animal models (especially large animal models) can provide insight into the structure and function of tissues when no human tissue data are available. Characterization of cartilages from animal models is also useful for the translation of cartilage therapies by showing whether an animal model is appropriate for use in preclinical studies, as further

R. C. Nordberg et al., *Tissue Engineering the Cartilages of the Body*, Synthesis Lectures on Tissue Engineering, https://doi.org/10.1007/978-3-032-09105-5_2

elaborated in Chaps. 4 and 5. In this chapter, human cartilage anatomy will be described with the inclusion of human characterization data where possible. Animal characterization data will also be included and clearly indicated to supplement the discussion and to fill in gaps in understanding when little or no human data are available.

It should be noted that functional properties reported in the literature can vary depending on the biochemical assay performed or the testing parameters used in mechanical testing. For example, for collagen determination, assays may use hydroxyproline or a specific collagen type as its standard, and the data can be normalized to the wet weight, dry wet, or total protein. Thus, care should be taken when comparing biochemistry data across various studies or cartilages. For mechanical testing, various testing methods and mechanical models to describe cartilage are presented in Chap. 4. In this chapter, an effort has been made to identify studies that use the same testing methods so that the properties presented can be compared across different cartilages. Unless otherwise denoted, tensile properties (i.e., Young's modulus and ultimate tensile strength (UTS)) were obtained from uniaxial strain-to-failure testing at a rate of 1% deformation per second. Also, unless otherwise denoted, creep indentation data were analyzed using a semi-analytical, semi-numerical, linear biphasic model and finite element analysis [140]. When values reported in this chapter were obtained via different testing protocols, they are clearly denoted.

2.1 Knee Cartilage

Within the knee, there are two articulating surfaces referred to as the tibiofemoral and patellofemoral joints. Within these joints, multiple anatomical structures are covered with articular hyaline cartilage. On the femoral side of the knee, hyaline cartilage covers the medial and lateral femoral condyles, the trochlear groove, and the posterior surface of the patella. Hyaline cartilage also covers the medial and lateral aspects of the tibial plateau. In addition to articular hyaline cartilage, two fibrocartilaginous menisci aid in load transmission and enhance the congruence of the tibiofemoral joint. The overall structure of the knee is illustrated in Fig. 2.1.

2.1.1 Knee Articular Hyaline Cartilage

Articular hyaline cartilage is a major structural element of the knee and is highly susceptible to injury and degeneration. It has been reported that over 60% of patients undergoing knee arthroscopy were documented to have chondral lesions [141]. Additionally, it has been estimated that about 14 million people in the United States, or approximately 4% of the population, suffer from symptomatic knee OA [142]. Therefore, cartilage injury and degeneration within the knee represent a significant health burden.

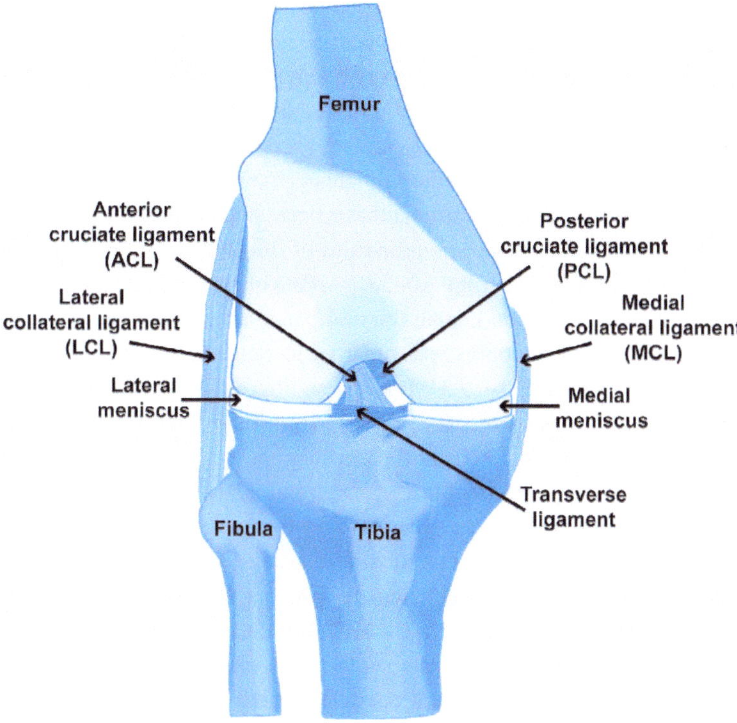

Fig. 2.1 Knee joint anatomy. Articular cartilage covers the distal end of the femur and the proximal end of the tibia. The knee meniscus is located between these two surfaces. Additionally, the patella (not depicted) has a cartilaginous surface that articulates with the trochlear groove of the femoral surface. The posterior cruciate ligament, anterior cruciate ligament, lateral collateral ligament, and medial collateral ligament restrict movement to maintain normal joint functionality. Reproduced with permission [41]

Biochemically, hyaline cartilages of the knee consist primarily of collagen and GAG. Collagen content of the human medial condyle was reported to be 17.2–21.6% by wet weight and 47.1–78.4% by dry weight [31]. Cartilage properties in other compartments of the knee (e.g., lateral condyle and trochlear groove) are also available in various animal models. In the sheep, collagen by dry weight was reported to be 67.5–81.7% in the medial condyle, 70.2–77.9% in the lateral condyle, 63.6–72.7% in the trochlear groove, and 73.4–76.8% in the patella [5]. In the bovine model, collagen content by wet weight was reported to range from 14.4–16.1% in the condyles, 16.5–16.6% in the tibial plateau, 17.6–18.4% in the trochlear groove, and 13.3–14.5% in the patella [143]. In a porcine model, it was estimated that 91.1% of total protein is collagen2A1, and the second most abundant collagen was collagen3A1, consisting of 4.4% of total protein [144]. GAG content of the human medial condyle was reported to be 5.0–6.8% by wet weight and 18.2–20.9% by dry weight [31]. In the sheep, GAG by dry weight was reported to be 16.3–17.6% in the medial condyle, 13.6–19.0% in the lateral condyle, 15.3–16.8% in the trochlear groove, and 16.5–16.8% in the patella [5]. In the bovine model, GAG content by wet weight was reported to range from 12.7–13.6% in the condyles, 14.5–14.6% in the tibial plateau, 13.4–14.5% in the trochlear groove, and 11.5–12.6% in the patella [143]. Altogether, knee cartilage has biochemical variability among species and joint regions, which should be noted for generating location-specific implants.

Mechanical properties of knee cartilage have been studied in the human and multiple animal models. Specifically, the tensile properties of full-thickness human medial condyle samples were reported to range from 7.0–30.7 MPa Young's modulus and 4.2–18.9 MPa UTS [31]. In a separate study on human femoral condylar cartilage, when tested at a rate of 5 mm/min, the Young's modulus was reported to range from 6.6–130 MPa and UTS was reported to range from 3.7–19.3 MPa, depending on depth, collagen fiber orientation, sex, and proximity to degenerate regions [145]. In the sheep, Young's modulus was reported to range from 5.8–18.6 MPa in the medial condyle, 14.5–19.3 MPa in the lateral condyle, 15.2–28.5 MPa in the trochlear groove, and 12.1–13.0 MPa in the patella [5]. UTS was reported to range from 3.3–10.1 MPa in the medial condyle, 9.9–13.3 MPa in the lateral condyle, 9.0–17.3 MPa in the trochlear groove, and 7.7–9.0 MPa in the patella of the sheep [5]. In the bovine model, tensile Young's modulus was reported to range from 6.1–6.7 MPa in the condyles, 6.7–7.1 MPa in the tibial plateau, 6.3–7.5 MPa in the trochlear groove, and 4.8 MPa in the patella [143]. An interspecies characterization study that used creep indentation of the lateral condyle, medial condyle, and patellar groove to obtain aggregate modulus values reported 530–701 kPa in the human, 472–899 kPa in the cow, 555–904 kPa in the dog, 522–815 kPa in the monkey, and 516–714 kPa in the rabbit [146]. In another study, aggregate modulus was reported to range from 379–777 kPa in different locations within the human medial condyle [31]. In the human patella, one study reported an average equilibrium modulus of 790 kPa using a confined compression testing setup, although this study included samples that were both healthy and arthritic [147]. In the sheep, compressive aggregate modulus was reported to range from 183–292 kPa in the medial condyle, 283–299 kPa in the lateral condyle, 269–364 kPa in the trochlear groove, and 193–297 kPa in the patella [5]. In the bovine model, the compressive aggregate modulus was reported to range from 354–383 kPa in the condyles, 400–419 kPa in the tibial plateau, 387–391 kPa in the trochlear groove, and 326–348 kPa in the patella [143]. Tribologically, the coefficient of friction has been reported to range from 0.22–0.26 on the human medial condyle [31]. Overall, knee hyaline cartilage is relatively well characterized compared to other joints, but additional characterization of healthy human tissue in locations such as the patella would provide valuable insights to the field.

2.1.2 Knee Meniscus

The knee menisci are fibrocartilaginous structures that function by transferring load and improving congruence between the femoral condyle and the tibial plateau. The knee contains both medial and lateral menisci that are stabilized by a network of ligamentous structures (Fig. 2.2). The meniscus has regional variation between the inner and outer regions of the structure, known as the white and red regions, respectively. Specifically, the white inner region is characterized by a lack of vasculature, while the red outer region is vascularized. There is also a transition region between the inner and outer meniscus, known as the red-white region, which has sparse presence of blood vessels. Clinically, it has been estimated that in the United States, approximately 850,000 meniscal-related orthopaedic surgeries are conducted per year [43].

Biochemically, the meniscus consists of approximately 70% water and 30% organic matter. Of the organic matter, approximately 75% is comprised of collagen, 15% is comprised of GAGs, and the rest is comprised of other proteins such as elastin, adhesion glycoproteins, and DNA [41]. Specifically, in the minipig, it was reported that by wet weight, the meniscus is 23.9–31.3% collagen and 1.2–2.6% GAG [43]. GAG content has been reported to consist of approximately 40% chondroitin-6-sulfate, 10–20%

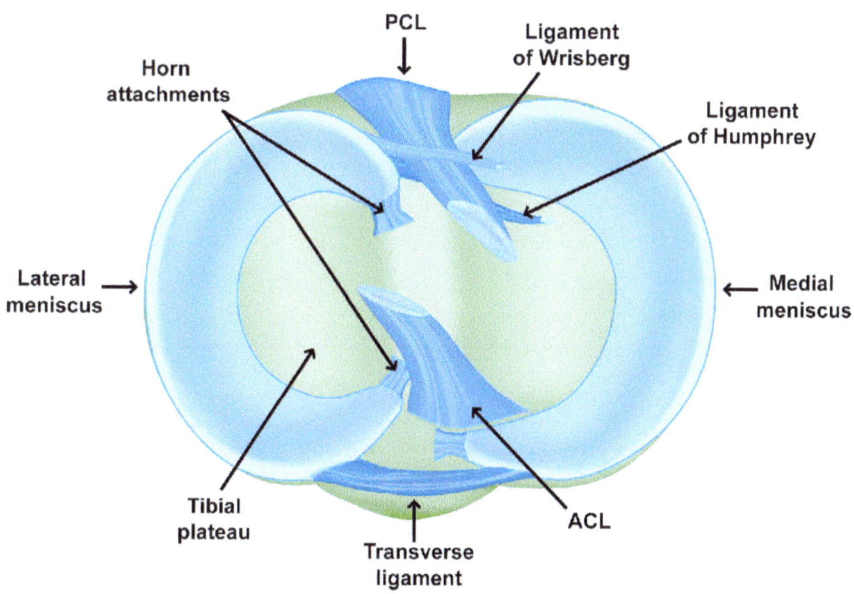

Fig. 2.2 Medial and lateral menisci anatomy. The menisci are attached to the tibial plateau via horn attachments. The transverse ligament, anterior cruciate ligament (ACL), posterior cruciate ligament (PCL), and ligaments of Wrisberg and Humphrey aid in restricting movement within this joint. Reproduced with permission [41]

chondroitin-4-sulfate, 20–30% dermatan sulfate, and 15% keratan sulfate [41]. Historically, it has been thought that the collagen content of the red region of the meniscus consists of approximately 99% type I collagen, and the collagen content of the white region consists of approximately 40% type I collagen and 60% type II collagen [41]. However, a recent study using a bottom-up proteomic technique in the minipig meniscus reported that total protein of the red region consisted of 61.3% collagen1A1, 30.2% collagen1A2, 5.8% collagen3A1, 0.6% collagen2A1, and that total protein of the white region consisted of 50.7% collagen1A1, 21.5% collagen1A2, 19.7% collagen3A1, and 4.8% collagen2A1 [144]. Therefore, the presence of type II collagen in the white region may be lower than originally thought. Due to these discrepancies, collagen subtyping characterizations of the meniscus in the human and other animal models would be of interest to the field.

Mechanically, the meniscus generally has higher tensile properties and lower compressive properties than knee articular hyaline cartilage. For example, in an interspecies characterization of the medial meniscus, the aggregate modulus values ranged from 90 to 160 kPa in the human, 110–210 kPa in the cow, 190–260 kPa in the dog, 120–500 kPa in the rabbit, and 150–180 kPa in the baboon [148], which are lower values than reported in a similar study of articular hyaline cartilage [146]. In the minipig, it was reported that circumferential tensile Young's modulus values ranged from 78.4 to 116.2 MPa and radial tensile Young's modulus ranged from 2.5–10.9 MPa, while compressive aggregate modulus ranged from 157 to 287 kPa [43]. In another study, the compressive aggregate modulus of the femoral aspect of the medial porcine meniscus was reported to be 270 kPa, 170 kPa, and 140 kPa in the anterior, central, and posterior segments, respectively [149]. In the same study, similar magnitudes were reported in the tibial aspect of the meniscus and on the lateral meniscus with values ranging from 130 to 270 kPa in all locations tested [149]. Altogether, the fibrocartilaginous nature of the knee meniscus gives rise to relatively high tensile properties and low compressive properties compared to the hyaline cartilages of the knee.

2.2 Hip Cartilage

Anatomically, cartilage covers the acetabulum and the femoral head of the hip joint (Fig. 2.3). This joint is a diarthrodial ball-and-socket joint with the acetabulum acting as the socket and the femoral head acting as the ball. The ligamentum teres is an interarticular ligament that connects these two structures. While the acetabular and femoral head cartilages are primarily hyaline in nature, the acetabulum also has a fibrocartilaginous ridge around the periphery of the structure called the labrum, which acts to help stabilize the joint. Degeneration is more common in the inferior, non-weight-bearing region of the femoral head, which may be due to the disparity in compressive properties of the articulating tissues [150]. Clinically, the incidence of symptomatic hip OA has been estimated

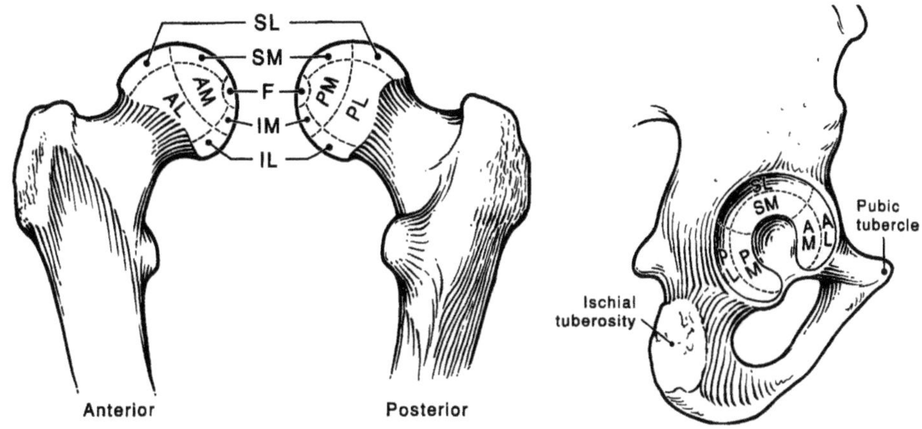

Fig. 2.3 Hip cartilage anatomy. Cartilage covers the femoral head and acetabulum of the hip. Anatomical regions are depicted here where A = anterior, S = superior, P = posterior, I = inferior, M = medial, L = lateral, and F = fovea. Reproduced with permission [150]

to be approximately 88 out of 100,000 persons [151], with over a million hip replacement procedures conducted worldwide each year [152].

Biochemical data of healthy human hip cartilage are relatively sparse compared to the knee, but animal data are available. In a porcine characterization study, both the acetabular and femoral cartilages were reported to be approximately 60% collagen by dry weight and 20% GAG by dry weight in the adult [153]. In the porcine model, collagen content in the femoral head normalized to total protein was reported to be 89.87% collagen2A1, 3.21% collagen3A1, and 2.17% collagen1A1 [144]. Mechanically, in the bovine model, the acetabular tensile modulus (tested in uniaxial tensile testing at a rate of 0.5 mm/min) was reported to range from 60.9 to 88.4 MPa, and the aggregate modulus measured via confined compression was reported to range from 139 to 174 kPa [154]. The tensile modulus (tested in uniaxial tensile testing at a rate of 10 mm/min) in the human acetabular labrum was reported to be 66.4 MPa, but these samples were obtained from patients undergoing hip surgery and may not be representative of the healthy population [155]. The tensile modulus (tested in uniaxial tensile testing at a rate of 0.5 mm/min) of human femoral head cartilage was reported to be 121.1 MPa in osteonecrotic specimens and 145.4 MPa in non-osteonecrotic specimens [156]. In an interspecies compressive characterization of hip cartilage, aggregate modulus values in the cow ranged from 290 to 480 kPa in the acetabulum and 220–550 kPa in the femoral head, in the dog ranged from 760 to 900 kPa in the acetabulum and 480–1,050 kPa in the femoral head, in the baboon ranged from 950 to 1210 kPa in the acetabulum and 710–1110 kPa in the femoral head, and in the human ranged from 1180 to 1320 kPa in the acetabulum and 820–1440 kPa in the femoral head [140]. In a separate study, the human acetabulum aggregate modulus ranged from 1072 to 1424 kPa, and the femoral head aggregate modulus ranged from 679 to 1816 kPa [150]. Further characterization studies, particularly those assessing biochemical content, would be valuable contributions to the field.

2.3 Ankle Cartilage

The tibiotalar (i.e., ankle) joint is located at the junction of the distal tibia, distal fibula, and proximal talus (Fig. 2.4). The ends of these three bones are covered with hyaline cartilage, but degenerative changes in the cartilage are most often located on the talus. Compared to the knee and hip, where OA often develops as a function of age, OA of the ankle generally develops after a traumatic injury. OA of the ankle is less frequent than OA of the knee, afflicting approximately 1% of the population [157].

Functional characterizations of the ankle joint suggest that a mechanical mismatch in articulating surfaces may contribute to degeneration within this joint. Biochemical data of human ankle cartilage are scarce; however, in the bovine model, the collagen by wet

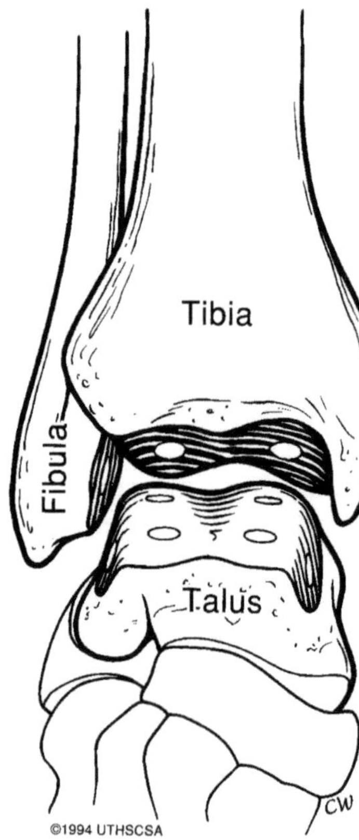

Fig. 2.4 Tibiotalar joint anatomy. Cartilage covers the three articulating surfaces of the tibiotalar joint. Specifically, the cartilages of the distal tibia and distal fibula articulate with the cartilage of the proximal talus. Reproduced with permission [158]

weight was reported to be 20.5%, 14.4%, and 8.4% in the tibia, fibula, and talus, respectively, and GAG content by wet weight was reported to be 18.0%, 10.8%, and 10.1% in the tibia, fibula, and talus, respectively [159]. Mechanically, in the human, the aggregate modulus of the distal tibia has been reported to range from 940 to 1340 kPa [158]. The fibular cartilage had an aggregate modulus of 1140 kPa. The talus, on the other hand, had aggregate modulus values that ranged from 920 to 1210 kPa. There was a notable mismatch in the mechanical properties of the posterior aspects of the ankle; the posterior tibia had an average aggregate modulus of 1250 kPa, whereas the posterior aspects of the talus had an average aggregate modulus of 920 kPa. In bovine tissue, mechanical disparities between the tibia and talus were observed [158]. For this animal model, the aggregate modulus values were 370 kPa, 242 kPa, and 255 kPa in the tibia, fibula, and talus, respectively [159]. Whereas tensile properties for human ankle cartilage remain to be characterized, a bovine study reported Young's modulus values of 4.82 MPa, 3.86 MPa, and 1.7 MPa in the tibia, fibula, and talus, respectively [159]. Given that both human and bovine tibia cartilage exhibited higher values in their mechanical properties than talus cartilage, researchers have postulated that this difference may account for the prevalence of cartilage degeneration in the talus.

2.4 Temporomandibular Joint (TMJ) Cartilage

The TMJ (Fig. 2.5) is a bilateral ginglymodiarthrodial joint (i.e., a joint that functions in both rotation and translation) that facilitates jaw movement essential in activities such as talking and eating. It is a two-compartment joint, where a fibrocartilaginous disc (sometimes referred to as a meniscus) articulates inferiorly with the mandibular condyle and superiorly with the glenoid fossa and articular eminence of the temporal bone. During mastication, the mandibular condyle is the primary moving structure of the joint, whereas the fossa-eminence structure remains stationary. The disc facilitates the transmission of tensile and compressive loads across the joint and improves the congruence between the mandibular condyle and fossa-eminence. A meta-analysis of TMJ disorder epidemiology has reported that close to a third of the population is symptomatic [160]. TMJ disorders include conditions such as internal derangement (i.e., misalignment or loss of normal disc movement), anterior displacement of the disc with or without reduction, disc thinning, and disc perforation. Treatment of these disorders includes palliative options, such as corticosteroid injections, or major surgical interventions, such as discectomy. However, no treatment options currently exist that restore the cartilage function to the joint. This section will describe the three cartilaginous structures within the TMJ (i.e., disc, mandibular condyle, and temporal fossa), and the functional characteristics of these structures.

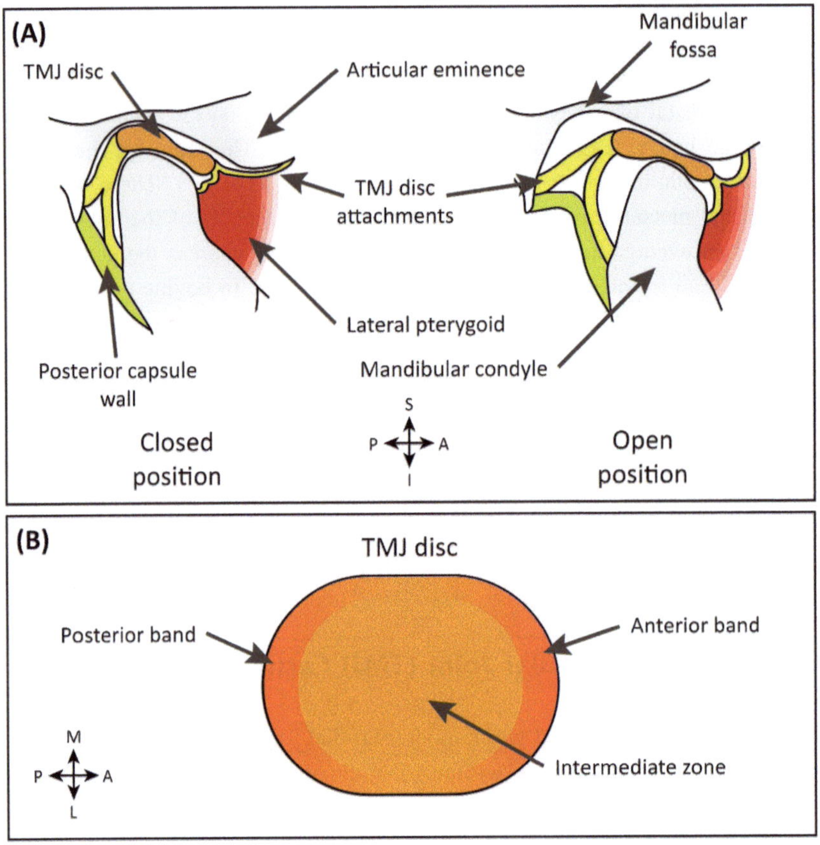

Fig. 2.5 Temporomandibular joint (TMJ) disc anatomy. **a** The TMJ disc is located between the mandibular condyle and the fossa-eminence and articulates with these surfaces as the jaw opens and closes. The disc is held in place by attachments, which surround the entirety of the disc and create two separate compartments within the joint. **b** The disc has bands in both the anterior and posterior positions, and an intermediate zone comprises the middle portion. Abbreviations: A = anterior, I = inferior; L = lateral, M = medial, P = posterior; S = superior. Reproduced with permission [115]

2.4.1 TMJ Disc

The TMJ disc is a fibrocartilaginous structure that has been described as biconcave and elliptical in shape, being longer in the mediolateral direction than the anteroposterior direction [38]. The disc has three distinct morphological zones (i.e., the posterior band, the intermediate zone, and the anterior band) that run along the mediolateral axis. The posterior and anterior bands are thicker than the intermediate zone, acting as rails to maintain positioning as the joint moves during mastication. In the human, the dimensions of the TMJ are approximately 19 mm in the mediolateral axis by 13 mm in the anteroposterior;

the thickness ranges from approximately 4 mm in the posterior band to approximately 1 mm in the intermediate zone [38].

Biochemically, the disc is primarily water and collagen. The disc hydration has been estimated to range from 66 to 80% [161], with collagen consisting of 83–96% of the dry weight [38]. Specifically, in the human, collagen by wet weight ranged from 20.3 to 23.1%, and GAG by wet weight ranged from 0.5 to 0.8% [162], Similarly, in the minipig, which is a commonly used animal model for TMJ disc repair, collagen by wet weight was reported to range from 26.7–27.8%, and GAG by wet weight was reported to range from 0.7 to 1.0% [163]. Additionally, in the minipig, collagen1A1 and collagen1A2 were reported to comprise 92.2% of total protein, while collagen2A1 and collagen3A1 comprised 1.1% and 4.1% of total protein, respectively [144]. GAGs compose approximately 5% by dry weight, with chondroitin-4-sulfate and chondroitin-6-sulfate composing 74–79% of the GAG content [38]. The rest of the GAG content consists of keratan sulfate chains, hyaluronan, and dermatan sulfate proteoglycans (e.g., decorin and biglycan). There is also a small amount (i.e., 1–2% by dry weight) of elastin reported in the disc, which likely contributes to restoring the shape of the tissue after loading [38].

Mechanically, the TMJ disc is substantially softer in compression than in tension. Tensile modulus values of the TMJ disc are dependent on the anatomical location from which the sample is taken (e.g., posterior band vs. intermediate zone) and testing orientation. In the human, tensile Young's modulus values were reported to range from 37.2 to 51.7 MPa in the anteroposterior direction and 9.5–25.5 MPa in the mediolateral direction [162]. In the minipig, tensile Young's modulus values were reported to range from 3.3 to 31.0 MPa in the anteroposterior direction and 1.8–26.3 in the mediolateral direction [163]. In terms of compressive properties, in the human, in a stress relaxation test at 20% strain, instantaneous modulus ranged from 287 to 2823 kPa, relaxation modulus ranged from 17.8–71.9 kPa, and coefficient of viscosity ranged from 2.9 to 20.0 MPa [162]. The minipig TMJ disc has been reported to have very similar compressive properties to the human TMJ disc; in a stress relaxation test at 20% strain, instantaneous modulus ranged from 216 to 1540 kPa, relaxation modulus ranged from 20.5 to 57.5 kPa, and coefficient of viscosity ranged from 5.9 to 23.0 MPa [163]. While the TMJ discs of other animal models, such as the farm pig, cow, goat, and rabbit, have also been characterized [162], the dimensions, biochemical properties, and mechanical properties of the TMJ disc in the minipig closely resemble those of the human TMJ disc, thereby supporting its use as the leading animal model for TMJ disc research.

2.4.2 Mandibular Condyle Cartilage

In the inferior compartment of the TMJ, the disc articulates with the mandibular condyle. Mandibular cartilage has distinct regions, including a fibrous zone, proliferative zone, mature zone, and hypertrophic zone, as depicted in Fig. 2.6 [164]. While the biochemical composition of the mandibular condyle has not been fully elucidated, it has been reported to contain type I collagen in the fibrous superficial zone rather than type II collagen [165]. In tension, the mandibular condyle's properties have been shown to be dependent on location and orientation. Specifically, in a porcine model, the Young's modulus of the mandibular condyle was reported to be 24.0 MPa in the anteroposterior direction and 10.1 MPa in the mediolateral direction, with additional variation in these values depending on which location the tissue was isolated from (e.g., medial, central, or lateral region of the condyle) [166]. In compression, using a porcine model, the apparent aggregate modulus of the mandibular condyle was reported to be 62 kPa, and the intrinsic aggregate modulus was reported to be 51 kPa, where "apparent and intrinsic properties of the tissue were defined in the literature to distinguish the properties of the tissue with and without the osmotic effects, respectively." [167]. It has been noted that the compressive properties of the mandibular condyle are relatively low compared to other articular cartilages. The relatively low aggregate modulus value has been attributed to a lack of proteoglycan content in the superficial zone of the cartilage [167]. Additionally, it has been proposed that the TMJ transfers compressive load through a trampoline-like mechanism, which uses the tensile properties of the collagen fibers in the disc to transmit loads, thereby shielding the mandibular condyle from direct compressive loading [168].

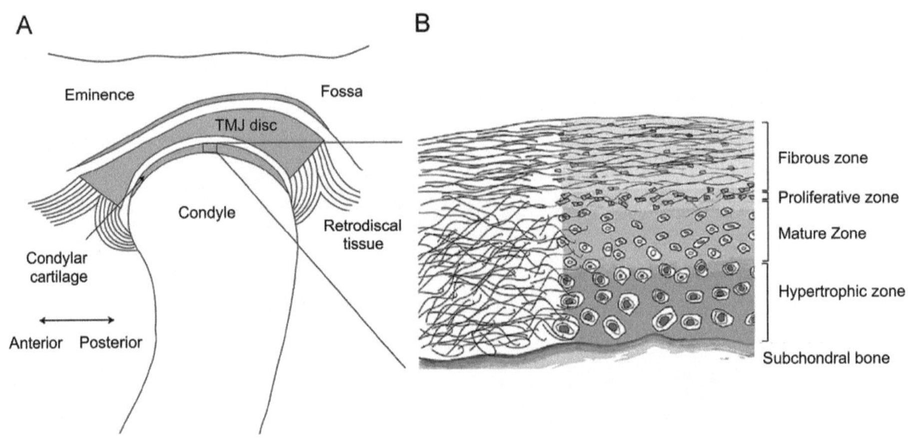

Fig. 2.6 Mandibular condyle anatomy. **a** The mandibular condyle is located below the disc within the temporomandibular joint (TMJ), where cartilage is depicted in dark gray. **b** Condylar cartilage displays distinct zonal organization with four distinct regions (i.e., the fibrous zone, the proliferative zone, the mature zone, and the hypertrophic zone), each of which has distinct matrix fiber organization and cellular compositions. Reproduced with permission [164]

2.4.3 Temporal Fossa Cartilage

In the superior compartment of the TMJ, the disc articulates with the temporal fossa. Similar to the mandibular condyle, the temporal fossa has distinct depth-dependent zones, including a fibrous layer, a prechondroblastic zone, a cartilaginous zone, and a zone of mineralization [12]. While there have been relatively few studies on the functional properties of the temporal fossa, in a porcine model, it has been reported that this cartilage has an aggregate modulus ranging from 35.0 to 58.9 kPa, depending on the cartilage's location, with an average of 41.9 kPa [12]. Similar to the mandibular condyle, the temporal fossa exhibits low compressive properties, likely due to a fibrous superficial zone with limited proteoglycan content and a disc that transmits forces primarily via tension rather than compression.

2.5 Cartilages of the Spine

The spine has over 100 cartilaginous joints (e.g., an estimated 23 IVDs, 46 facet joints, 24 costovertebral joints, 20 costotransverse joints, 2 sacroiliac joints, an atlanto-axial joint, and an occipito-atlantal joint). Between each vertebra of the spine, an IVD and two facet joints make up a three-joint complex that works together to transmit loads across the spine and provide stability to the spine. At the base of the spine, the sacroiliac joint transmits loads from the spine to the pelvis. Of the joints of the spine, the IVD has received the most attention in literature due to its widespread acceptance as a contributor to back pain. There is also a growing appreciation for how the facet joints and sacroiliac joint contribute to back pain and disability. Thus, this section will focus on the structure and function of cartilages within these three joints. Other spinal structures that have cartilaginous components that lack thorough characterization include the atlas and axis at the base of the skull and the costovertebral and costotransverse joints that connect the ribs to the spine.

2.5.1 Intervertebral Disc (IVD)

The IVD is a soft, cylindrical tissue that resides between the vertebral bodies of the spine (Fig. 2.7). The two primary zones within the IVD are a fibrocartilaginous annulus fibrosus along the periphery of the disc and a gel-like nucleus pulposus at the center of the disc that is not cartilage. Additionally, a band of hyaline cartilage is present where the disc meets the boney vertebral body in a structure known as the end plate. The prevalence of disc degeneration is high, with reports of disc degeneration over the entire spine afflicting greater than 90% of the population over the age of 50 [169].

Developmentally, the annulus fibrosus arises from the mesoderm through mesenchymal condensation, in contrast to the nucleus pulposus, which arises from the notochord,

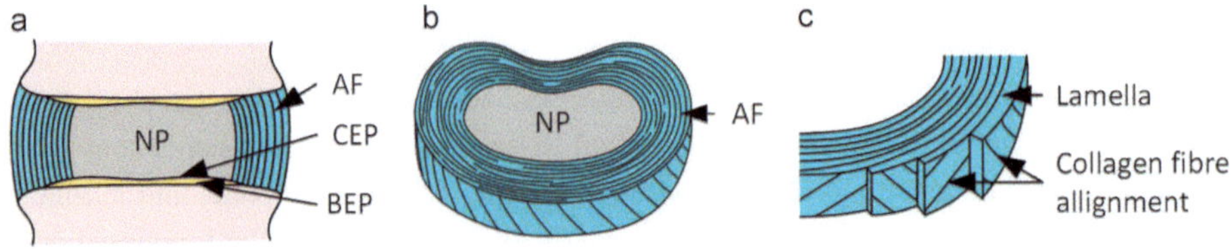

Fig. 2.7 Intervertebral disc (IVD) anatomy. Cross sections of the IVD in the **a** coronal plane, and **b** sliced transversely. The anulus fibrosus surrounds the nucleus pulposus in a concentric ring (i.e., lamellae) pattern. **c** The fiber (fibre) orientation rotates by approximately 60° in alternating lamellae. Abbreviations: AF = anulus fibrosus, CEP = cartilaginous endplate, BEP = bony endplate, NP = nucleus pulposus. Reproduced with permission [170]

imparting each structure with distinct structural and functional characteristics. The nucleus pulposus is not cartilaginous in nature but, rather, is primarily a gel which is 50% proteoglycan by dry weight and 20% collagen by dry weight [171]. The annulus fibrosus is a fibrocartilaginous structure consisting of coaxial lamellae that form a ring around the loose ECM network of the nucleus pulposus (Fig. 2.7). Within each lamella, the collagen fibers are aligned, and this orientation alternates in adjacent lamellae by rotating the orientation by approximately 60°, thereby providing a cross-hatching collagen fiber organization throughout the annulus [170]. In terms of biochemical organization of the annulus fibrosus, the inner lamellae have a higher concentration of type II collagen and GAGs, while the outer lamellae have higher concentrations of type I collagen and elastin. These lamellae may have distinct functions because the outer type I collagen-rich lamellae connect directly into the vertebral bone, while the inner type II collagen-rich lamellae blend into the hyaline cartilage end plates. In the minipig, it was reported that total protein of the annulus fibrosus consisted of 63.7% collagen1A1, 28.5% collagen1A2, 4.2% collagen3A1, and 1.2% collagen2A1 [144]. The end plate is composed of predominantly type II collagen, similar to other hyaline cartilages [170].

Generally, the mechanical properties of the IVD are not well characterized. However, it has been reported that in confined compression, the nucleus pulposus has a compressive aggregate modulus of approximately 1.0 MPa [172], but in unconfined compression, the equilibrium modulus was 5.4 kPa [173], demonstrating the need for the annulus fibrosus to confine the nucleus pulposus to impart its compressive properties. The tensile modulus of the annulus fibrosus has been reported to vary substantially depending on orientation and topographical location. Specifically, in a review that compiled data across multiple mechanical testing studies, the tensile modulus was estimated to be approximately 183 MPa along the axis of the fibers, 16 MPa circumferentially, and 0.3 MPa radially [170]. Additional studies that study the mechanical properties of the IVD are needed.

2.5.2 Facet Joint Cartilage

Between each two vertebrae of the spine, two facet joints are located dorsal to the IVD (Fig. 2.8) [174]. The facet supports spinal motion and stability, and, in conjunction with the IVD, transmits loads along the spine. The structure of the facet consists of two articular hyaline cartilage surfaces (a superior surface and an inferior surface) surrounded by a synovium and fibrous joint capsule. The facet joints are highly susceptible to degeneration; an interspecies characterization of facet cartilage study reported pathology on 100% of facet surfaces of the human facet tissue from 154 joints isolated from donors 41–80 years of age [175]. Of these, 72% of the surfaces examined were categorized as grade 4 degeneration with severe cartilage loss. Other studies have estimated that facet arthrosis affects 57% of adults by age 30 and 100% of adults by age 60 [176]. While procedures such as nerve blocks and neurotomies are able to temporarily manage pain, there is no treatment available that restores the structure and function to these joints [177].

Due to the high degree of degeneration in human facet tissue, characterization of healthy human facet cartilage is difficult to accomplish. However, in recent years, several characterization studies have been conducted to provide insight into the structure–function relationships of facet joint cartilage in animal models [175, 179–181] and arthritic human tissue [175, 182]. The architecture of facet cartilage is that of articular hyaline cartilage, with a superficial, middle, and deep zone. In the minipig, the facet joint is approximately 65.8–82.1% collagen by dry weight, and 9.0–16.4% GAG by dry weight [181]. Additionally, the type II to type I collagen ratio is consistent with other articular hyaline cartilages; 87.20% of the total protein has been reported to be type II collagen, and 4.16% has been reported to be type I collagen [144]. The compressive properties of facet cartilage are generally lower than other articular hyaline cartilage, such as that of the knee. For example, in the minipig, it has been reported that aggregate modulus values

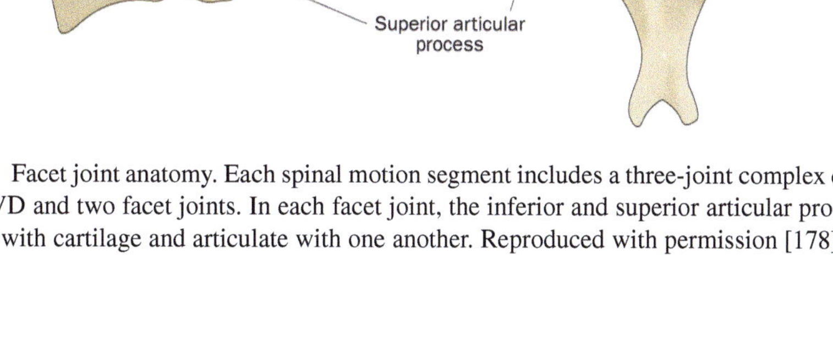

Fig. 2.8 Facet joint anatomy. Each spinal motion segment includes a three-joint complex consisting of the IVD and two facet joints. In each facet joint, the inferior and superior articular processes are covered with cartilage and articulate with one another. Reproduced with permission [178]

range from 135 to 262 kPa [181]. Tensile Young's modulus values were reported to range from 6.7 to 20.3 MPa in the minipig [181]. The coefficient of friction of Yucatan minipig facet joints was reported to range from 0.34 to 0.52, which is higher than values reported in the knee (ranging from 0.27 to 0.29) in the same study [180]. Although facet cartilage is hyaline in nature, it exhibits lower compressive properties and a higher coefficient of friction than knee hyaline cartilage, potentially making its functional properties easier to recapitulate through tissue engineering.

2.5.3 Sacroiliac Joint Cartilage

Bilateral sacroiliac joints transmit loads from the sacral spine to the ilium of the pelvis (Fig. 2.9a). The sacroiliac joint is estimated to be the locus of pain in 14–22% of lower back pain cases [183, 184]. The range of motion of this joint is small; it is estimated to have approximately 3° of movement in flexion–extension, 1.5° of movement in axial rotation, and 0.8° of movement in lateral bending [185]. As depicted in Fig. 2.9b, the structure of the sacroiliac joint consists of a ligamentous region and a cartilaginous region, and it has been proposed that the thick ligamentous region acts in tension in a trampoline-like mechanism to shield the cartilaginous region from significant compressive loading [186].

Sacral cartilage and iliac cartilage have distinct functional characteristics. There have been few characterization studies of sacroiliac cartilage, and no functional characterization of human sacroiliac cartilage. Generally, iliac cartilage is considered to be more fibrous in nature, while the sacral cartilage is more hyaline in nature [186]. In the Yucatan minipig, the biochemical composition by dry weight was reported to include 12.2% and 5.2% GAG in the sacral and iliac cartilages, respectively, and 44.2% and 45.6% collagen in the sacral and iliac cartilages, respectively [186]. In terms of tensile properties, sacral cartilage has been reported to have a Young's modulus of 13.6 MPa and 7.1 MPa in the cranial-caudal axis and dorsal–ventral axis in the minipig (corresponding to the anterior–posterior axis and cranial-caudal axis in the human), respectively [186]. Iliac cartilage has been reported to have a Young's modulus of 2.5 MPa and 3.5 MPa in the cranial-caudal axis and dorsal–ventral axis, respectively. Compressive properties were also dependent on anatomical location, with an aggregate modulus of 133 kPa and 51 kPa in the sacral and iliac cartilages, respectively [186]. Sacroiliac cartilage is not smooth as it is in a typical diarthrodial joint, but, rather, it has ridges that deepen with age. The sacroiliac cartilage has been quantified to have significantly higher surface roughness than cartilage of the knee, and a coefficient of friction ranging from 0.34 to 0.37 [180]. Overall, the functional disparity between sacral and iliac cartilages is significant, warranting further investigation to characterize the tissues to uncover the underlying causes of this disparity.

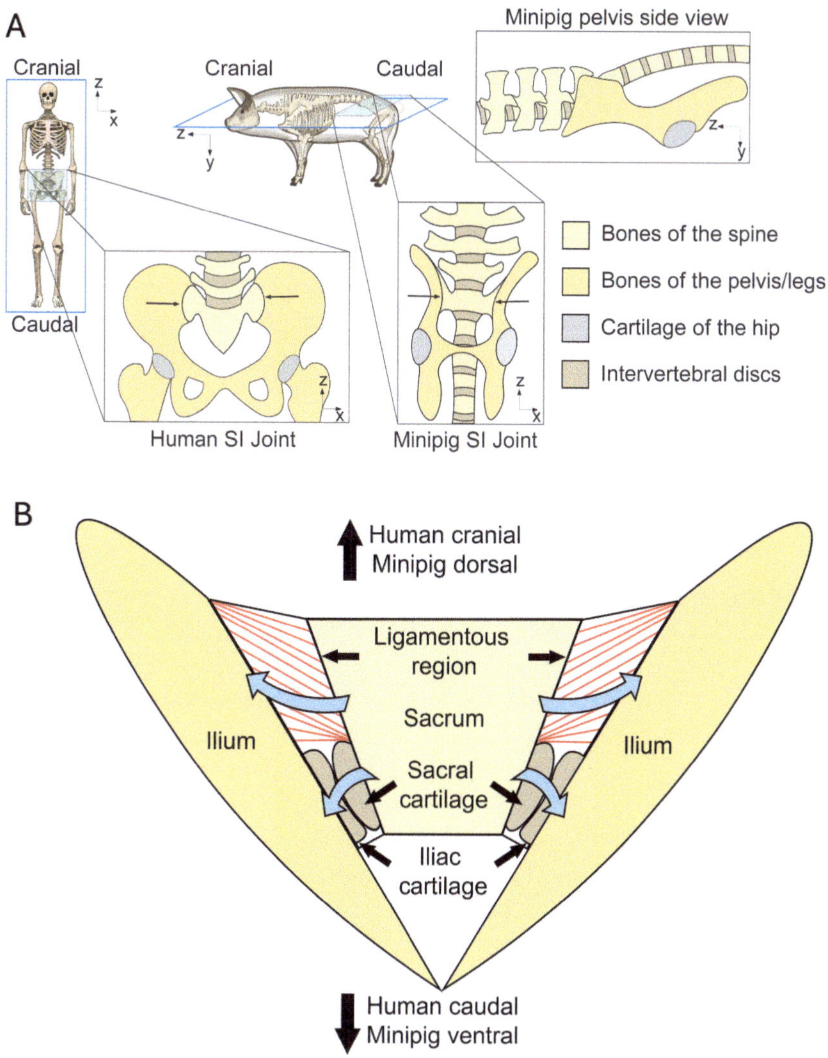

Fig. 2.9 Sacroiliac joint anatomy. **a** The location of the sacroiliac joint (black arrows) is shown in both the human and minipig. Despite differences in spinal orientation due to bipedal versus quadrupedal posture, the sacroiliac joint anatomy is relatively conserved between species. **b** Within the joint, both humans and minipigs possess a thick ligamentous region that works in conjunction with the sacroiliac cartilage to transmit loads from the sacrum to the ilium. Reproduced with permission [186]

2.6 Other Diarthrodial Joints

It should be noted that beyond the diarthrodial joints discussed in this chapter, there are numerous other diarthrodial joints within the human body. For example, the hands and feet have many joints with limited characterization data. Of these, the first metatarsophalangeal joint has been characterized for compressive properties with aggregate modulus values ranging from 630 to 1340 kPa [187]. Additionally, the second metatarsal intermediate cuneiform joint has been characterized for compressive properties with aggregate modulus values ranging from 880 to 1200 kPa [188]. Other major diarthrodial joints include the shoulder and elbow. In the canine shoulder, the mean humeral cartilage aggregate modulus was reported at 710 kPa, and the mean glenoid cartilage aggregate modulus was reported at 670 kPa [189]. In the human elbow, aggregate modulus values were reported to range from 624 to 899 kPa depending on anatomical location within the joint [190]. Overall, more characterization studies are needed in these joints and other diarthrodial joints not listed here.

2.7 Non-articulating Cartilages

Beyond being a major structural component of joints, cartilage is also present in numerous non-articulating structures of the body, such as the outer ear, nose, ribs, trachea, and other structures. While, for joints, the main function of cartilage is to transmit loads and provide a low-friction surface for articulation, the cartilages of non-joint structures have different functional requirements. Non-articulating cartilages primarily serve as flexible structural elements with defined geometries. This section will describe what is known about the structure and function of non-articulating cartilages. Beyond the structures that fall into the traditional cartilage classifications (hyaline cartilage, fibrocartilage, and elastic cartilage), this section will also describe a new type of cartilage termed lipocartilage that was recently described to be a major structural element of some craniofacial structures in mammalian species, including preliminary evidence in the developing human.

2.7.1 External Ear Cartilage

The structure of the external ear (i.e., auricle) consists of elastic cartilage surrounded by perichondrium and skin and has a complex architecture as depicted in Fig. 2.10. Reconstructing the auricle is of clinical relevance, particularly in cases of congenital abnormalities such as microtia. The prevalence of microtia has been reported to range from 0.83 to 17.4 of 10,000 births and varies substantially depending on demographic factors [191]. Due to the psychosocial issues that may arise from congenital microtia, there is a need to develop tissue engineering solutions to repair the auricle.

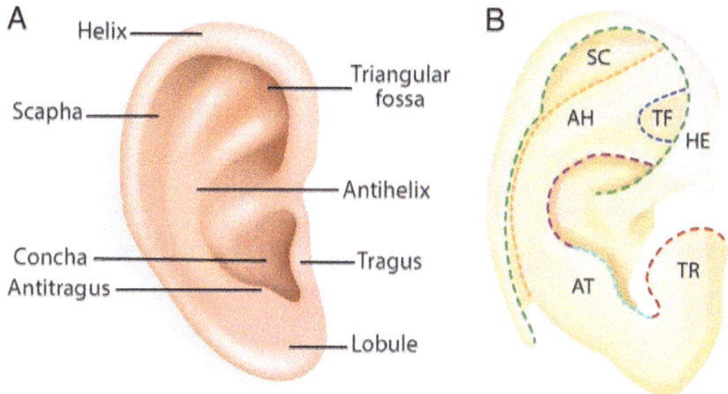

Fig. 2.10 Ear cartilage anatomy. **a** The outer ear has distinct regions, each with unique geometric features. **b** These regions are supported by underlying elastic cartilage structures. Abbreviations: AH = antihelix, AT = antitragus, HE = helix, SC = scapha, TF = triangular fossa, TR = tragus. Reproduced with permission [192]

The biochemical composition of ear cartilage is difficult to define from the literature due to inconsistent findings and reporting styles, and values range widely. Many studies in the auricle report hydroxyproline content rather than collagen content. While the total collagen content can then be estimated from the hydroxyproline measurements using a conversion factor, different researchers may use different collagen standards or conversion factors, making the data across many papers difficult to compare. Hydroxyproline was reported to range from 73.3 to 97.9 nmol/mg by wet weight in young adult donors and 55.2–65.1 nmol/mg by wet weight in old adult patients [193]. One characterization of human concha found that collagen only consists of approximately 0.6% of the dry weight [194]. In contrast, in porcine auricular cartilage, it has been reported that the percentages of collagens from total protein measured using mass spectrometry were 34.4% collagen1A1, 18.1% collagen1A2, 40.1% collagen2A1, and 3.8% collagen3A1 [144]. Elastin content of the human ear concha was reported to be 14.1% by dry weight [194], but another report found elastin content to range from 12.2 to 20.7% by wet weight [193], meaning the per dry weight value is lower. In the human, sulfated GAG content was reported to range from 2.1 to 3.7% by wet weight in young adult donors and 1.1–1.5% in old adult donors [193]. In bovine ear, the GAG content by dry weight was reported to be approximately 22% [195]. Overall, the field would benefit from additional characterization studies that include biochemical content.

The mechanical properties of human auricular cartilage have been characterized in both compression and tension. Compressive Young's modulus values have been reported as 1.1 MPa, as quantified via nanoindentation in a study that examined cadaveric auricular cartilage isolated from the ear concha [194]. Another study loaded samples in compression to 300 g at a rate of 1 mm/s as a function of anatomical structure of the ear, and reported a compressive Young's modulus of 1.7 MPa for the tragus, 1.8 MPa for the antitragus, 2.1 MPa for the concha, 1.7 MPa for the antihelix, and 1.4 MPa for the helix [196]. Tensile properties have also been investigated, with one study (testing in uniaxial tension at a rate of 0.1 inch/min) reporting a human auricular cartilage Young's modulus of 5.1 MPa and a UTS of 2.2 MPa [197]. These values are in close agreement with another study, which reported a maximum tensile strength of 2.0 MPa and a tensile Young's modulus of 5.0 MPa of human auricular samples (testing in uniaxial tension at a rate of 50 mm/min) [198]. Overall, the mechanical characterization of the ear cartilage is relatively more consistent in comparison to the biochemical characterizations.

2.7.2 Nasal Cartilage

Nasal cartilages are hyaline in nature, and there are three main nasal cartilage substructures: the septum, the paired upper lateral cartilage (ULC), and paired lower lateral cartilage (LLC), as depicted in Fig. 2.11. The septum divides the nasal cavity into two cavities and provides midline structural support to the nose. The ULC is contiguous with the septum but forms the outer structure to the lateral wall of the nasal cavities. The LLC provides structure to the lateral wall of the tip of the nose and consists of the lateral, intermediary, and medial crura regions. The transition between the ULC and LLC is varied and can consist of overlapping cartilage, cartilage joined end-to-end, or no direct cartilage contact. Additionally, each nasal region has depth-dependent microstructure including superficial, intermediate, and central zones, with cells being less dense and more round in morphology in the central zones of the tissue. Functionally, the septum, ULC, and LLC cartilages work together to provide structure to the nose, which is essential for proper respiratory function.

Functional properties of nasal cartilage show characteristics that are in line with other hyaline cartilage types. For example, nasal cartilage is comprised of predominantly type II collagen (ranging from approximately 70–85% of total collagen depending on anatomical structure) [4], which is consistent with other hyaline cartilages. It also contains approximately 5–15% type I collagen and 1–7% type III collagen when normalized to total collagen [4]. The biochemical content also contains trace amounts of IX, X, and XI collagens, as well as GAGs and elastin. The collagen content of human nasal cartilage was estimated to range from 9.3 to 15.9% collagen by wet weight, while GAG content was estimated to range from 2.5 to 5.3% GAG by wet weight [4]. In terms of mechanical properties, the compressive aggregate modulus of human nasal tissue has been reported to range from 283 to 449 kPa in the septum, 171–191 kPa in the ULC, and 155–239 kPa in the LLC. In the same study, the tensile Young's modulus was reported to range from 4.6 to 10.1 MPa in the septum, 7.1–13.6 MPa in the ULC, and 4.1–6.8 MPa in the LLC. It is notable that, although the nose is a non-weight bearing structure, nose cartilage has compressive and tensile moduli comparable to those of weight-bearing cartilages such as facet joint cartilage and knee cartilage. While strides have been made toward elucidating the functional properties of nasal cartilages, more characterization is needed for different donor ages, donor sexes, and relevant animal models.

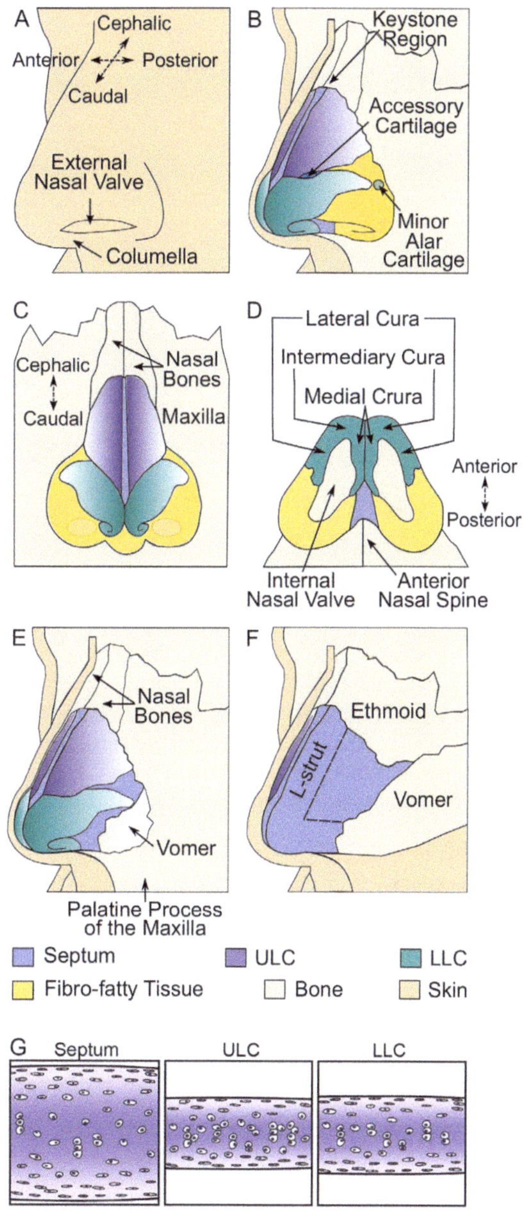

Fig. 2.11 Nasal cartilage anatomy. **a** Anatomical orientations of the nose. Major cartilaginous structures within the nose include the septum, upper lateral cartilage (ULC), lower lateral cartilage (LLC), minor alar cartilage, and accessory cartilage. Here, these structures are depicted from **b** an angled frontal view, **c** a frontal view, **d** a worm's-eye view, **e** an angled frontal view with fibro-fatty tissue removed, and **f** an angled frontal view of the septal cartilage. **g** Cross sections of the major cartilage types show that the cellular morphology and organization of each cartilage are unique. Reproduced with permission [4]

2.7.3 Rib Cartilage

Costal (i.e., rib) cartilage, which is hyaline in nature, is categorized as true (i.e., sternal) ribs, false (i.e., asternal) ribs, or floating ribs (Fig. 2.12). True ribs connect directly to the sternum, false ribs connect to the sternum through the true ribs, and floating ribs do

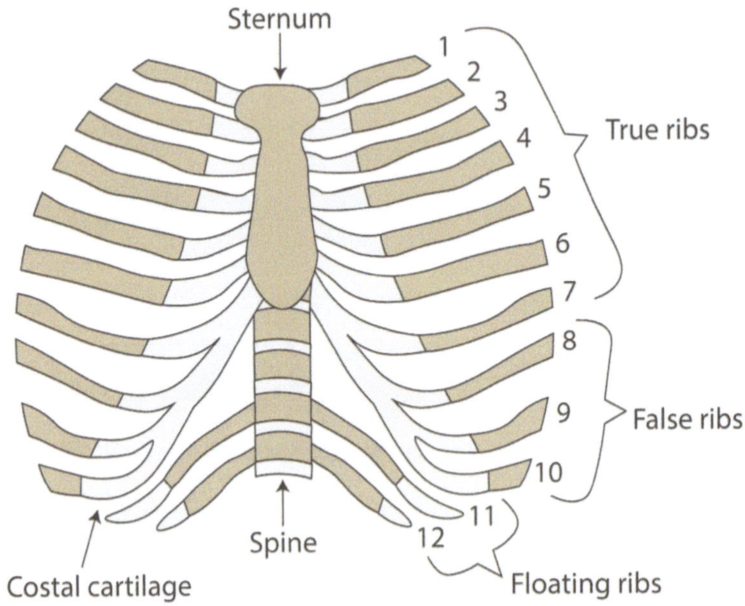

Fig. 2.12 Costal cartilage anatomy. The human ribcage typically contains 12 pairs of ribs. The costal cartilage of ribs 1–7 (i.e., true ribs) connect directly to the sternum. The costal cartilage of ribs 8–10 (i.e., false ribs) connect to the sternum through the costal cartilage of rib 7. Ribs 11 and 12 (i.e., floating ribs) have no anterior attachment to the sternum but end in a free-floating cartilage tip. All ribs articulate posteriorly with the spine (not depicted in ribs 1–10) through costovertebral joints and costotransverse joints

not connect to the sternum. Costal cartilage is known to undergo partial ossification as a function of age [199]. It has been used as a graft tissue source for reconstructive surgeries for indications such as TMJ, nasal, or tracheal repair [5]. Clinically, the rib has been used for reconstructive surgeries for indications such as ear [200] or nasal [201] repair. Costal cartilage also may need to be reconstructed, such as in traumatic injuries from automotive crashes. Costal cartilage fractures account for approximately 7.8% of blunt chest trauma injuries [202]. To date, there have been limited characterization studies on costal cartilage functional properties. Semi-quantitative studies based on histology have identified collagen and mineralization within the samples [203], as well as the presence of type II collagen [204]. In the sheep, GAG by dry weight was reported to be 22.6%, and collagen by dry weight was reported to range from 28.9 to 32.3% [5]. The proteomic composition of minipig costal cartilage has been reported to be 6.4% collagen1A1, 0.3% collagen1A2, and 80.9% collagen2A1 when normalized to total protein [144]. In the sheep, compressive aggregate modulus was reported to range from 691 to 1151 kPa. Tensile Young's modulus was reported to range from 4.0 to 8.1 MPa depending on the location and orientation of the test [5]. Additional characterization of human costal cartilage is necessary, with particular attention to changes associated with aging and mineralization.

2.7.4 Tracheal Cartilage

The trachea is responsible for maintaining a patent airway during respiration. Tracheal reconstruction can be necessary due to conditions such as traumatic injury, congenital abnormalities, and cancer [205]. The structure of the trachea includes 18–22 cartilaginous horse-shoe-shaped rings that are closed on the posterior side by the trachealis muscle (Fig. 2.13). The cartilaginous rings are held together longitudinally by the annular ligament. While the biochemical composition of tracheal cartilage has yet to be fully elucidated [57], the cartilage within the trachea is considered to be hyaline in nature and is predominantly type II collagen [205]. It has been reported that tracheal cartilage consists of approximately 19% GAG by dry weight in the pig [206]. The tensile Young's modulus of human tracheal tissue has been reported to range from 5.1 to 39.6 MPa (tested in uniaxial tension at a strain rate of 0.5 mm/min) [207]. It was also reported that the tensile Young's modulus values of the porcine tracheal cartilage were 5.6 MPa in the circumferential direction and 1.1 MPa in the longitudinal direction (tested in uniaxial tension at a strain rate of 0.5% strain/second), indicating the tracheal cartilage is an anisotropic tissue [206]. In terms of compressive properties, the aggregate modulus of porcine tracheal cartilage was reported to be 1300 kPa (tested in confined compression in approximately 5% increments of strain) [206]. There is significant potential for further characterization of tracheal cartilage, including comprehensive biochemical analysis and compressive testing of human tissue.

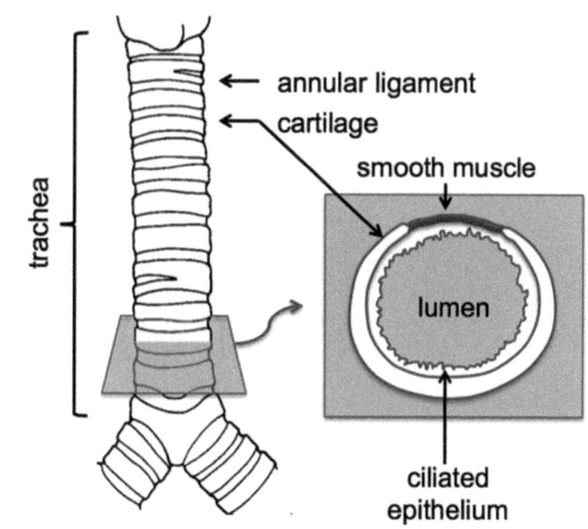

Fig. 2.13 Tracheal cartilage anatomy. The cartilage of the trachea forms horseshoe-shaped rings that are joined longitudinally through the annular ligament, with smooth muscle forming interconnecting bridges. Reprinted with permission from Boazak EM, Auguste DT. Trachea Mechanics for Tissue Engineering Design. *ACS Biomater Sci Eng* 2018:1272–84. Copyright 2018 American Chemical Society [205]

2.7.5 Growth Plate Cartilage

Growth plate cartilage is present in the developing long bones of children and serves as the template tissue that ossifies during long bone elongation. The growth plate has several distinct zones classified as the resting zone, proliferative zone, hypertrophic zone, and calcification zone (Fig. 2.14). Clinically, the growth plate is of significance because it is mechanically weaker than the surrounding bone and is prone to fracture and bony bar formation (i.e., a bridge of bone that spans the growth plate), which can lead to deformed or stunted limb growth in severe cases. It has been estimated that in the United States, approximately 350,000 children sustain growth plate injuries each year, particularly in the upper limbs, ankle, and phalanges [208, 209].

Characterizing the functional properties of the growth plate can be challenging due to its heterogeneous nature. In the bovine distal femur, hydroxyproline content was reported to be 33.0 μg/mg by dry weight, and GAG content was reported to be 27.1% by dry weight [210]. In the bovine distal femur growth plate using uniaxial tension at an approximate strain rate of 0.004/s, the average Young's modulus was reported to be 34.6 MPa, and the average UTS was reported to be 3.0 MPa [210]. Compression stress-relaxation experiments of the bovine growth plate using 20% strain reported aggregate modulus values ranging from 660 to 1290 kPa, depending on the location of the growth plate that

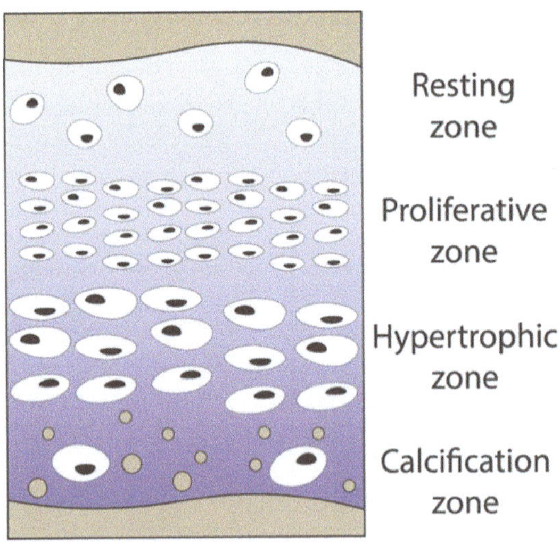

Fig. 2.14 Growth plate cartilage anatomy. The growth plate contains several zones with distinct cellular functions that support bone growth and elongation. Chondrocytes in the resting zone serve as a reservoir of cells for the growth plate. In the proliferative zone, chondrocytes actively divide to drive elongation. In the hypertrophic zone, chondrocytes enlarge, and the matrix begins to prepare for mineralization. The calcification zone is where the tissue mineralizes, and blood vessels invade to form bone tissue

was tested [211]. It has been reported that compressive properties are non-homogeneous among zones of the growth plate. In a study that used stress-relaxation tests with unconfined compression at 15% strain, compressive Young's modulus in the "transverse plane" was reported to be 8.7 MPa for the complete growth plate, 11.1 MPa for the reserve (i.e., resting) zone, 3.4 MPa for the proliferative zone, and 3.3 MPa for the hypertrophic zone [212]. In the same study, the "out-of-plane" compressive Young's modulus was reported to be 0.5 MPa for the complete growth plate, 0.5 MPa for the reserve zone, 0.3 MPa for the proliferative zone, and 0.27 MPa for the hypertrophic zone [212]. Another study used microindentation in a rabbit model and reported linear elastic Hertz reduced modulus values ranging from 344 to 526 kPa depending on the distance across the growth plate [213]. Overall, while there have been some characterization studies on the growth plate, additional characterization is needed using both bulk testing modalities and testing modalities that can probe the heterogeneities and anisotropy within the tissue. To enable direct comparisons with other cartilages, characterization studies are needed that employ testing modalities analogous to those used in previous cartilage characterization studies.

2.7.6 Lipocartilage

A novel cartilage type, termed lipocartilage, was recently described as a novel type of cartilage that does not fall within the traditional cartilage categorizations (i.e., hyaline cartilage, fibrocartilage, or elastic cartilage) [9]. In lipocartilage (Fig. 2.15), large intracellular lipid vacuoles are present within the tissue, and the cellular components compose a large portion of the tissue volume, in contrast to other cartilage types that are predominantly ECM. Lipocartilage has been identified in the nose, ear, larynx, and chest cartilages of certain mammalian species such as the mouse, rat, and bat. Additionally, preliminary evidence of this cartilage has been found in the human fetal thyroid, epiglottis, ear, and nasal cartilages [9]. Unlike adipose tissue, lipid vacuoles in lipocartilage do not respond to perturbations in caloric intake. Proteomic characterization identified the top ten most highly expressed proteins in ear lipocartilage to be 24.4% collagen1A1, 13.6% collagen1A2, 10.1% VIM (vimentin), 7.1% collagen3A1, 6.2% MYOC (myocilin), 4.5% collagen2A1, 3.7% collagen6A1, 2.8% Hist1h4h (histone H4), 2.5% ELN (elastin), and 2.5% Hist1h2bp (histone H2B) [9]. These results indicate that collagen and cell-associated proteins are the most abundant components. Mechanically, in the mouse model, the tensile Young's modulus values of lipocartilage were reported to be 3.3 MPa compared to 23.9 MPa and 23.9 MPa in the meniscus and rib, respectively [9]. Additionally, it was demonstrated that delipification of the lipocartilage increased the tensile modulus of the tissue, suggesting that the lipid vacuoles confer compliance to the tissue. Although further research is needed to elucidate the role of lipocartilage in human tissue morphogenesis and function, future studies on this emerging cartilage type may aid in the development of novel innovations in craniofacial tissue engineering and regenerative medicine.

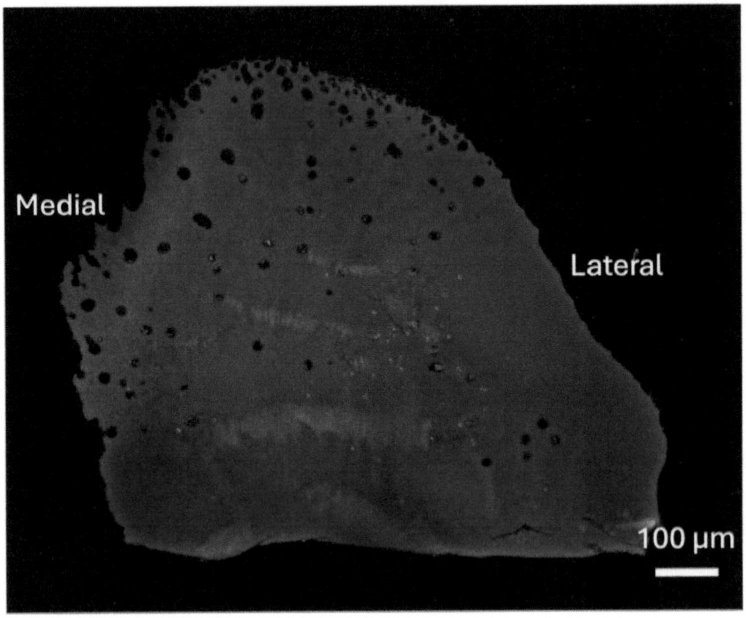

Fig. 2.15 Rodent ear lipocartilage anatomy. A fluorescent image of a whole-mount postnatal day 32 mouse ear lipocartilage stained with boron-dipyrromethene (BODIPY), with the medial side of the ear on the left and the lateral side on the right. BODIPY is lipophilic and localizes to lipid droplets throughout the tissue. Black spots observed throughout the tissue are fenestrae (i.e., holes), which are characteristic of rodent ear cartilage. Figure provided courtesy of Dr. Raul Ramos

Chapter concepts:

- Characterizing the functional (e.g., morphologic, biochemical, biomechanical) characteristics of native cartilage tissues can aid in (1) elucidating structure–function relationships and (2) developing design standards for tissue engineering cartilage implants.
- Human characterization data, while important for clinical implant development, can be difficult to collect due to scarcity of healthy tissue. Characterization of large animal tissues may provide additional insights into the structure and function of native cartilages.
- It can be difficult to directly compare values collected from different studies due to variability in testing protocols and reporting standards within the literature. It would be beneficial to develop standards for data collection and reporting.
- In the human, compressive aggregate modulus and tensile Young's modulus of knee articular cartilage have been reported to be 379–777 kPa and 7.0–30.7 MPa, respectively, depending on location and orientation [31, 146].

- In comparison to knee articular hyaline cartilage, the fibrocartilaginous knee meniscus has a lower compressive aggregate modulus (e.g., 90–160 kPa in the human) [148] but a higher tensile Young's modulus (e.g., 78.4–116.2 MPa in the minipig) [43].
- Further characterizations of other diarthrodial joints, such as the hip and ankle, are needed. For example, tensile properties of hip cartilage require further elucidation.
- In the TMJ, the human and minipig have been shown to have exceptionally similar functional properties and geometry.
- Cartilages of the spine, such as those of the IVD, facet joint, and sacroiliac joint, require much more characterization, particularly from healthy human tissues.
- Elastic cartilage of the ear has been reported to have a tensile Young's modulus of 5.0–5.1 MPa [197, 198], and a compressive Young's modulus of 1.1–2.1 MPa [194, 196], but the biochemical content is not well defined.
- Nasal cartilage has been reported to have a tensile Young's modulus ranging from 4.1 to 13.6 MPa and a compressive aggregate modulus ranging from 155 to 449 kPa in the human [4].
- Other non-articulating cartilages, such as those of the rib, trachea, and growth plate, remain undercharacterized and warrant further characterization studies.
- The tensile Young's modulus of lipocartilage in the mouse has been reported to be approximately 3.3 MPa [9]. However, additional research is needed to characterize compressive properties and establish the functional significance of this tissue in humans.

Cartilage Tissue Engineering Methods 3

Takumi Takahashi, Rachel C. Nordberg, Jerry C. Hu, and Kyriacos A. Athanasiou

Chapters 1 and 2 reviewed the structure and function of the various cartilage types and highlighted select native cartilages. This chapter turns to the tissue engineering of these cartilages with the goal of restoring or regenerating them throughout the body. The studies referenced earlier help define the design criteria for cartilage tissue engineering, including ECM composition, biomechanical properties such as compressive and tensile moduli, zonal organization, and the need for integration with surrounding tissues. These criteria serve as benchmarks for creating biomimetic neocartilage that not only resembles native cartilage in structure but also functions appropriately in its intended biological roles.

In this book, tissue engineering refers to the use of living cells in combination with signaling cues and/or scaffolds to restore, replace, or regenerate lost cartilage (Fig. 3.1). Two broad strategies are recognized. One involves the in vitro generation of neocartilage that mimics the structure and function of native cartilage. The other applies tissue engineering principles to stimulate the in vivo formation of neocartilage within the body. Both approaches are grounded in the three foundational pillars of tissue engineering: (1) cells, (2) a structural scaffold or organizational framework, and (3) biochemical and/or biomechanical signals. In practice, strategies are typically classified as either scaffold-based, in which biomaterials provide physical support and spatial organization, or scaffold-free, in which cells self-organize through externally applied energy or self-assemble in the absence of external forces to form neocartilage [214]. This framework emphasizes the tissue-forming intent of tissue engineering: not simply to modulate the biological environment but to actively form neocartilage in vitro or in vivo to ultimately serve in place of native cartilage.

R. C. Nordberg et al., *Tissue Engineering the Cartilages of the Body*, Synthesis Lectures on Tissue Engineering, https://doi.org/10.1007/978-3-032-09105-5_3

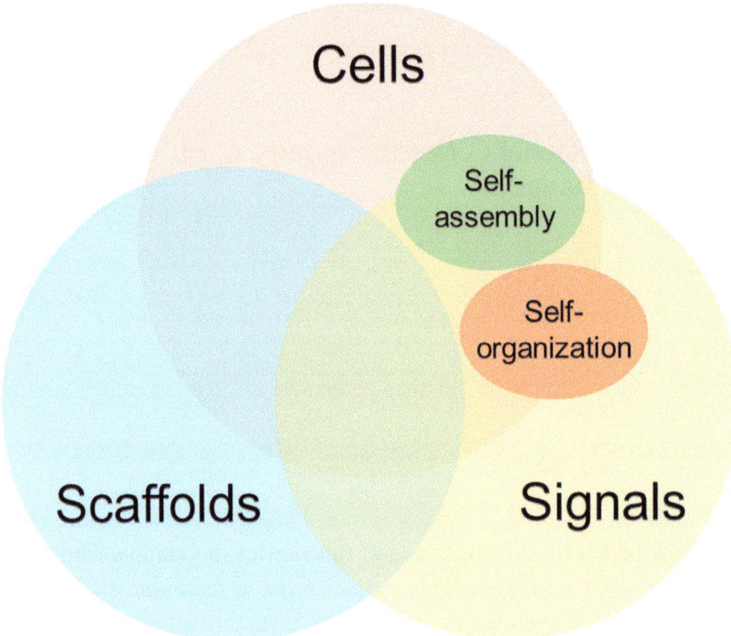

Fig. 3.1 Tissue engineering pillars. The three fundamental pillars of tissue engineering are cells, scaffolds, and signals. In scaffold-free approaches, self-assembly and self-organization are techniques that represent distinct strategies, differentiated by the absence or presence of externally applied energy, respectively [214]

Excluded from this definition are therapies that lack both cells and either a scaffold or ECM produced by the cells themselves. These include MSC injections without carriers, exosomes, extracellular vesicles (EVs), platelet-rich plasma (PRP), and bone marrow aspirate concentrate (BMAC) [215, 216]. These approaches primarily function through paracrine signaling or immunomodulation [217]. Many of these approaches offer temporary symptom relief, but multiple randomized clinical trials have failed to demonstrate consistent improvements in cartilage structure or function compared to placebo [218–224]. Also excluded are acellular implants and devices, such as the Collagen Meniscus Implant (CMI), Actifit polyurethane meniscus scaffold, NUBAC nucleus replacement for the IVD, and polyethylene and silicone implants used in auricular reconstruction [225–227]. While these implants may serve as defect fillers or scaffolds for host tissue infiltration, they are devoid of viable cells at the time of implantation, and tissue formation may be limited [228–230]. To summarize, these products are not categorized as tissue engineering because they either lack cells, a structural component to hold cells, or both.

Within this framework, the chapter proceeds through four sections that reflect the three pillars of tissue engineering: cell sources, scaffold-based approaches, scaffold-free approaches, and methods for modulating functional properties through biochemical and biomechanical cues.

3.1 Cell Sources

The selection of cell sources is a key decision in cartilage tissue engineering. Cells are responsible for synthesizing ECM, responding to environmental signals, and ultimately determining the structural and functional properties of neocartilage. The mechanical integrity and long-term performance of neocartilage are directly affected by the intrinsic characteristics of the chosen cell source. Key criteria include the cells' capacity to generate appropriate ECM components, maintain the ability to express the correct phenotype after expansion, expand to clinically relevant quantities, and integrate effectively with host tissue. Additional considerations include the accessibility of source tissue, donor-to-donor variability, immunogenicity, and alignment with regulatory and manufacturing requirements. Developing specific protocols for each tissue source for cell isolation, expansion, storage, and manipulation is essential to ensure consistent quality and reproducibility. Thus, identifying viable cell sources requires satisfying a complex set of scientific and translational criteria.

Cells may be utilized in autologous, allogeneic, or xenogeneic contexts. Autologous cells, which are derived from the same patient receiving the treatment, reduce the risk of immune rejection and have been successfully translated into clinical products such as matrix-induced autologous chondrocyte implantation (MACI). However, autologous approaches are limited by the need for multiple surgeries, one to harvest the tissue and another to implant the neocartilage, as well as donor site morbidity, limited availability of donor tissue, and variability in cell quality related to patient-specific factors such as age, sex, or comorbidities [3]. Moreover, because each autologous product is tailor-made for an individual patient and manufactured on-demand, these therapies tend to be expensive.

Allogeneic cells, which are sourced from donors other than the recipient, offer advantages in scalability, product standardization, and off-the-shelf availability (ready-made). These features make allogeneic products more suitable for widespread clinical use and more attractive for companies developing tissue-engineered cartilage products. However, the use of allogeneic cells introduces risks such as immune rejection and disease transmission, requiring stringent donor screening, sterility testing, and immunological assessment. Despite these challenges, allogeneic therapies are generally regarded as more commercially viable and cost-effective and are expected to play a larger role in future tissue-engineered cartilage applications [231].

Xenogeneic cells, which are derived from a different species, are not currently suitable for clinical use in humans because of the high risk of immune rejection and significant regulatory constraints. Instead, xenogeneic cells are typically used in preclinical models, such as in immunodeficient or immunosuppressed animals to test human cell-derived products. These models enable the direct evaluation of human tissue-engineered cartilage products in vivo while avoiding xenogeneic immune responses.

In addition to selecting among autologous, allogeneic, and xenogeneic cells, translational development often requires development of both human and animal products. Regulatory pathways for tissue-engineered cartilage products typically mandate pivotal large animal studies as part of obtaining regulatory approval for first-in-human trials [232, 233]. To comply with this requirement, researchers develop analogous cellular products (ACPs), which are species-specific counterparts to the human cell product, to test safety and efficacy in preclinical animal models [234]. For example, minipig cells have been used to make scaffold-free neocartilage and tested in minipig models for the treatment of the TMJ disc [235]. Parallel to the development of the ACP using minipig cells, human cells have been used to make scaffold-free neocartilage under the same protocol to demonstrate feasibility in developing the human product [236, 237]. These considerations, essential for successful translation of tissue-engineered cartilage products, are detailed in Chap. 6 along with recent comprehensive reviews [232, 238].

The four major categories of cells most commonly used in cartilage tissue engineering are chondrocytes, MSCs, embryonic stem cells (ESCs), induced pluripotent stem cells (iPSCs), and CPCs. A detailed definition and discussion of each, particularly those that have advanced to large animal studies and clinical trials, is provided below. Notably, most cell-based therapies have been developed in the context of hyaline cartilage regeneration, and this precedence has shaped the usage of these terms in the literature, which are often applied ambiguously and inconsistently. Each cell type brings distinct advantages. Chondrocytes remain the most clinically validated cell type for cartilage repair, particularly in autologous applications. MSCs are widely studied in clinical trials, offering broad tissue accessibility and scalability, particularly for allogeneic strategies. ESCs and iPSCs provide a large supply of cells but require stringent control of differentiation and raise concerns related to tumorigenicity. CPCs are promising for their cartilage specificity and expansion capacity, but they remain largely in preclinical stages of development. Ultimately, cell source selection must meet scientific and translational criteria that depend on the functionality of the cells and target cartilage type.

3.1.1 Chondrocytes

The term chondrocytes broadly refers to cells derived from cartilaginous tissues of the three classic types: hyaline cartilage, fibrocartilage, and elastic cartilage. Lipochondrocytes are derived from lipocartilage and are not discussed here because their use for tissue engineering is yet to be explored. Chondrocytes, in general, differ in functional properties based on numerous factors such as anatomical location of the tissue source, age, sex, disease, and methods of isolation, expansion, and cryopreservation. Thus, the tissue source of chondrocytes and methods of manipulation must be explicitly stated in tissue-engineered cartilage studies. The distinction between orthotopic and heterotopic use is also important. Orthotopic refers to the use of cells in their native anatomical context (e.g., using articular chondrocytes to repair articular hyaline cartilage), whereas heterotopic refers to use outside the native anatomical context (e.g., using nasal chondrocytes to repair articular hyaline cartilage) [239]. Here, the use of chondrocytes derived from articular, nasal, and costal cartilage is discussed, although other sources such as tracheal cartilage have been investigated to a more limited extent [240, 241].

Articular chondrocytes derived from hyaline cartilage of the knee have been the earliest adopted cell source for tissue-engineered cartilage. Articular chondrocytes obtained from non-weight-bearing regions of the knee are used orthotopically for hyaline cartilage repair [239]. These cells are clinically applied in autologous chondrocyte implantation (ACI) and MACI, as well as in autologous chondrocyte cell sheets [231, 242, 243]. Although articular chondrocytes are the most well characterized and used clinically, their autologous use is limited by factors such as donor site morbidity, variability in tissue quality, and limited tissue source [244].

Articular chondrocytes obtained from deceased donors or surgical remains serve as allogeneic alternatives to autologous sources. Human adult articular chondrocytes have been expanded to high passages (up to passage 11) to produce scaffold-free neocartilage for potential allogeneic use [236]. However, studies have shown that articular chondrocytes obtained from donors less than 13 years of age form neocartilage with 100-fold higher proteoglycan content than neocartilage formed from cells of older donors [244]. Thus, the use of young donors may be advantageous for allogeneic approaches. For example, human infant and juvenile articular chondrocytes have been sourced from surgical remains from the hip (hip dysplasia and Perthes disease) and supernumerary digits (polydactyly), circumventing the availability of healthy deceased donors as well as regulatory restrictions in the use of cadavers in certain regions [245–247]. In particular, human polydactyly-derived articular chondrocytes have been tested in first-in-human allogeneic transplantations of chondrocyte cell sheets to treat cartilage defects associated with knee OA, demonstrating safety of the allogeneic approach [248]. Articular chondrocytes, whether autologous or allogeneic, remain a viable cell source for cartilage tissue engineering, with juvenile cells offering enhanced ECM production and translational potential.

Nasal chondrocytes, derived from the hyaline cartilage of the nasal septum, have been used in both orthotopic and heterotopic applications. These cells have been isolated and expanded from multiple species, including human, ovine, and caprine nasal cartilages, and have demonstrated high proliferative capacity and the ability to form hyaline-like cartilage after multiple passages [249–251]. In preclinical studies, ovine nasal chondrocytes have been used to tissue engineer tracheal cartilage, resulting in constructs with biochemical and biomechanical properties similar to tracheal cartilage formed from native tracheal chondrocytes [240, 252]. In clinical settings, human nasal chondrocytes have been used to engineer neocartilage for the reconstruction of nasal alar lobules following tumor resection

[253]. Human and caprine nasal chondrocytes have also been used to engineer neocartilage for implantation into articular cartilage defects of the knee. In particular, human nasal chondrocytes have been tested in a first-in-human trial for the treatment of focal cartilage defects of the knee as well as larger defects in patients with knee OA [254, 255]. A phase II clinical trial for the treatment of focal cartilage defects subsequently demonstrated significant clinical improvement and hyaline-like cartilage regeneration following implantation [256]. Nasal chondrocytes are considered highly promising for clinical applications because of their accessibility for autologous use, expansion potential, and ability to integrate into both articular and non-articular cartilage environments.

Costal chondrocytes (sometimes called costochondral cells), derived from hyaline cartilage in the rib, have been increasingly studied for tissue-engineered cartilage applications, primarily in heterotopic contexts. These cells are relatively easy to harvest in large quantities, demonstrate high proliferative capacities, and form robust hyaline-like neocartilage [237, 257, 258]. Costal chondrocytes have been used clinically for pellet-type ACI [259, 260]. They have also been evaluated in scaffold-free neocartilage systems for potential allogeneic use. Both minipig and human costal chondrocytes form mechanically robust neocartilage in vitro [234, 237, 258]. These cells have been validated across multiple species, including rodent, porcine, caprine, ovine, and human, and continue to be investigated for use in knee articular cartilage, nasal cartilage, and TMJ disc [4, 237, 257, 258, 261–263]. A large animal study in the Yucatan minipig demonstrated healing of TMJ disc perforations using neocartilage formed from minipig costal chondrocytes, supporting the use of hyaline-like neocartilage for fibrocartilage repair [235]. Beyond joint and craniofacial applications, a study used costal chondrocytes in auricular tissue engineering, although for this particular study, costal chondrocytes showed calcification compared to neocartilage formed from nasal, articular, and auricular chondrocytes [264]. To a limited extent, costal chondrocytes have been used for tissue engineering tracheal cartilage [265, 266]. These diverse applications highlight the translational versatility of costal chondrocytes and support their continued development for both autologous and allogeneic use in cartilage tissue engineering.

Fibrochondrocytes (chondrocytes derived from fibrocartilage such as the knee meniscus, annulus fibrosus of the IVD, or TMJ disc) have been investigated for the tissue engineering of fibrocartilage. These cells typically express higher levels of type I collagen and lower levels of proteoglycans and type II collagen compared to chondrocytes derived from hyaline cartilage. Fibrochondrocytes from the knee meniscus have been isolated from multiple species, including leporine, bovine, ovine, and human for tissue engineering the knee menisci [267–271]. In particular, a 1:1 co-culture of articular chondrocytes and fibrochondrocytes from the knee menisci have been used to make wedge-shaped menisci without scaffolds [271]. Other scaffold-free approaches have used fibrochondrocytes from the knee meniscus, but testing in animal models has been limited compared to the use of MSCs [272, 273]. Fibrochondrocytes derived from the annulus fibrosus of the IVD (sometimes referred to as annulus fibrosus cells) have been used in preclinical leporine models [274, 275]. Fibrochondrocytes derived from the TMJ disc have been characterized as fibrochondrocytes but may be limited in accessibility and the capacity to be expanded for use in cartilage tissue engineering [276, 277]. These fibrochondrocytes are best suited for fibrocartilaginous tissue repair rather than hyaline cartilage regeneration and are primarily used for such purposes.

Chondrocytes, derived from the elastic cartilage of the external ear (sometimes referred to as auricular chondrocytes), are the most widely studied and clinically applied elastic cartilage-derived chondrocytes. Auricular chondrocytes, compared to knee articular chondrocytes, can be obtained with minimal donor-site morbidity, yield approximately twice as many cells per gram of tissue, and exhibit faster proliferation in monolayer culture [278]. Auricular chondrocytes naturally produce an ECM containing elastin in addition to type II collagen and GAGs, giving them shape-retaining flexibility. Human auricular chondrocytes, passaged up to P5, have been shown to form native auricular cartilage-like neocartilage in vitro [279]. In first-in-human studies, a two-stage procedure was applied to generate shape-specific ears for microtia repair [280–283]. From autologous auricular cartilage from the microtia affected ear, elastic chondrocytes were isolated and cultured and then injected into the patient's lower abdomen. After 6 months, neocartilage surrounded by perichondrium was collected and sculpted into an ear framework and transplanted to the microtia ear. Although auricular chondrocytes have been investigated for use in repairing other cartilage types, such as nasal and tracheal cartilages, their primary application remains in auricular reconstruction [284–286]. Other locations of elastic cartilage, such as the laryngeal epiglottis, larynx, and Eustachian tube, have been less explored for tissue engineering purposes. Among the elastic cartilage types studied, ear cartilage remains by far the most accessible, well-characterized, and clinically translated tissue source.

In summary, chondrocytes derived from hyaline cartilage, fibrocartilage, and elastic cartilage represent a diverse set of cell sources with distinct phenotypic profiles, matrix-forming capacities, and tissue-specific applications. Hyaline cartilage-derived chondrocytes, such as those from articular, nasal, and costal cartilages, are best suited for regenerating hyaline cartilage. Fibrochondrocytes and elastic chondrocytes are less frequently used in tissue engineering studies, yet they provide a distinct phenotype that may be useful for application in fibrocartilage and elastic cartilage repair, respectively. The selection of chondrocyte tissue source should be guided by tissue accessibility, the intended transplantation site, functional demands of the target tissue, and the translational context of the therapy being developed.

3.1.2 Mesenchymal Stromal/Stem Cells (MSCs)

MSCs are a heterogeneous population of multipotent progenitor cells capable of differentiating into various mesenchymal lineages, including chondrogenic, osteogenic, and adipogenic lineages. In the context of cartilage tissue engineering, MSCs have been extensively studied as a cell source because of their accessibility, capacity for in vitro expansion, and ability to undergo chondrogenic differentiation under appropriate biochemical and biomechanical cues. In the context of this book, MSCs do not include cartilage-resident progenitor cells, which we refer to as CPCs and discuss in Sect. 3.1.4. Here, we provide a discussion of MSCs that are not native to cartilage but, instead, reside in various other tissues but have nonetheless been investigated for cartilage tissue engineering.

MSCs can be derived from bone marrow, adipose tissue, synovium, umbilical cord, and other tissues [287]. Among these, bone marrow-derived MSCs (BM-MSCs) are the most extensively studied and have demonstrated strong chondrogenic potential. However, their yield is limited, and harvesting is invasive [288]. BM-MSCs from the iliac crest and vertebral body have shown greater chondrogenic capacity than those from the femoral head, indicating that anatomical origin influences MSC function [289]. MSC-like progenitor cells may also be isolated from non-cartilaginous tissues such as dermal tissues, and such dermis-isolated cells have been applied to scaffold-free cartilage tissue engineering and discussed elsewhere [290–292].

For knee cartilage defects, BM-MSCs have been used clinically with mixed results. A randomized pilot study comparing matrix-induced autologous BM-MSC implantation (m-AMI) and the chondrocyte-based MACI for full-thickness cartilage lesions found superior clinical and functional outcomes with m-AMI at 24 months, with no adverse events or graft failures [293]. In another study, autologous BM-MSCs delivered on platelet-rich fibrin sealant resulted in clinical and radiographic improvement in patients with full-thickness femoral condyle defects over 12 months, supporting the utility of BM-MSCs in this scaffold-based approach [294]. In a separate trial, allogeneic BM-MSCs were co-implanted with autologous chondrons obtained from debrided cartilage tissue [295, 296]. DNA short tandem repeat analysis of biopsy samples at 12 months confirmed that the regenerated cartilage consisted solely of autologous chondrocytes, with no contribution from the allogeneic MSCs. The BM-MSCs, comprising 80% or 90% of the implanted cell population, potentially contributed to cartilage regeneration through trophic effects. While most clinical studies to date have used BM-MSCs in intra-articular injections (non-tissue-engineering approaches), recent meta-analyses suggest that such injections do not lead to functional improvement in the knee, although modest reductions in pain may occur [223, 297, 298]. Thus, BM-MSCs may need to be used within a tissue engineering approach to have any therapeutic impact with the benefit largely driven by transient trophic effects.

For repairing the knee meniscus, implantation of autologous BM-MSCs seeded on a collagen scaffold into avascular meniscal tears was found to be safe and led to clinical improvement in three of five patients over 24 months in a first-in-human study; two patients experienced retears that required meniscectomies [299]. Other clinical studies demonstrated the safety of BM-MSCs for augmenting meniscal repairs but not in a tissue engineering context. In a randomized, double-blinded, controlled trial, intra-articular injection of allogeneic human BM-MSCs after partial meniscectomy in 35 patients led to increased meniscal volumes in only three patients but with no major safety concerns observed over 2 years [300]. While these studies suggest the feasibility and safety of BM-MSCs for cartilage repair, most clinical trials in this space have been limited by small sample sizes and effectiveness of treatments, underscoring the need for further development in the field.

Adipose-derived MSCs (AD-MSCs) are another widely studied population because of the abundance and accessibility of adipose tissue (fat tissue). AD-MSCs can be obtained through minimally invasive lipoaspiration procedures and expanded efficiently in vitro [301]. While they are generally more proliferative than BM-MSCs, their intrinsic chondrogenic potential is considered to be lower [302]. However, with appropriate chondrogenic differentiation, AD-MSCs can form hyaline-like cartilage and have been explored in preclinical and clinical trials for knee focal defects and knee OA. Preclinical studies in leporine and ovine models have used AD-MSCs and scaffolds to treat knee cartilage defects and knee OA, but further clinical translation has been limited [303, 304]. In a prospective study, culture-expanded autologous AD-MSCs embedded in a biodegradable matrix resulted in significant clinical, functional, and radiological improvement at 3-year follow-up in patients with symptomatic full-thickness knee cartilage defects [305]. In a review of clinical trials up to 2022, over 180 clinical trials using AD-MSCs were registered with clinicaltrials.gov, but a majority used intra-articular injections to treat knee OA, without the incorporation of additional biomaterials [306]. In a phase II/III study, cell-based intra-articular injections, including those with AD-MSCs, showed no improvement over corticosteroid injections [307]. While AD-MSCs represent an accessible cell source, their effective use may require further advances in tissue engineering strategies.

Synovium-derived MSCs (SY-MSCs) have gained attention for their high chondrogenic potential and intra-articular origin. These cells exhibit superior cartilage ECM production compared to BM-MSCs and AD-MSCs in several studies and are thought to retain a degree of tissue-specific memory that enhances their ability to repair joint cartilage [308, 309]. SY-MSCs can be harvested from the synovial membrane through arthroscopy, making them particularly relevant for orthopaedic applications. Their localization within the joint also suggests potential for reduced immunogenicity and improved engraftment. In a porcine preclinical model, transplantation of neocartilage from SY-MSCs resulted in superior cartilage regeneration compared to neocartilage made from dental pulp-derived MSCs. In a first-in-human study, surgical repair of complex degenerative medial meniscal tears followed by autologous SY-MSC transplantation resulted in significant improvements in clinical outcomes over 2 years without serious adverse events [310]. In another prospective clinical study, meniscus repair combined with autologous SY-MSC transplantation in patients with degenerative flap and radial medial meniscus tears led to significant improvements in patient-reported outcomes and arthroscopic findings at 12 months, with no major adverse events [311]. Together, SY-MSCs have demonstrated potential use in both hyaline cartilage and fibrocartilage repair, especially in augmenting meniscal repairs.

Umbilical cord-derived MSCs (UC-MSCs), typically isolated from Wharton's jelly or umbilical cord blood, offer an allogeneic source of MSCs with high proliferative capacity and immunomodulatory properties [312]. The only approved UC-MSC product (Cartistem) has been studied in preclinical cartilage repair models and clinical trials [313, 314]. Cartistem, an allogeneic UC-MSC product containing HA hydrogel, is approved in South Korea for the treatment of knee cartilage defects. In the United States, a phase I/IIa clinical trial for knee cartilage defects was completed in 2017, and in Japan, a phase III clinical trial for patients with moderate to severe knee OA began in 2023 [315, 316]. In 2024, Cartistem failed to receive approval in South Korea for the additional orthopaedic indications of treating ankle joint talar cartilage and osteochondral defects [317]. Despite rapid clinical progress with UC-MSCs, their therapeutic efficacy remains to be investigated.

In summary, MSCs derived from sources such as bone marrow, adipose tissue, synovium, and umbilical cord offer an accessible and versatile cell source for cartilage tissue engineering. Their capacity for chondrogenic differentiation under defined conditions, along with immunomodulatory properties and scalability, makes them promising candidates for both autologous and allogeneic use. Moving forward, well-designed clinical studies are still needed to determine whether any MSC source, in combination with a given tissue engineering strategy, can deliver consistent therapeutic benefit and surpass current standard-of-care treatments.

3.1.3 Embryonic Stem Cells and Induced Pluripotent Stem Cells (ESCs/iPSCs)

ESCs and iPSCs are pluripotent cells capable of differentiating into any cell type of the three germ layers, including mesodermal and neural crest lineages relevant for cartilages. Their ability to self-renew, i.e., undergo cell division and maintain pluripotency, makes them promising sources for generating large quantities of cells. In cartilage tissue engineering, both ESCs and iPSCs have been investigated as alternative cell sources [318–320]. While ESCs are inherently limited to allogeneic use, iPSCs derived from either the patient or a donor have been studied for autologous and allogeneic applications, respectively. Two modes of differentiation are typically employed. The first involves the direct differentiation of ESCs or iPSCs into chondrocyte-like cells [319–324]. The second approach is through the generation of lineage-specific, self-renewing, progenitor populations that are expanded prior to being differentiated into chondrocyte-like cells [325–328]. Although such progenitors as derived in the laboratory are sometimes termed MSCs or CPCs in the literature, they are referred to here as ESC-derived or iPSC-derived progenitors. The following paragraphs describe the use of ESCs and then iPSCs in cartilage tissue engineering, reflecting the historical order of their discovery.

Stable ESC lines, derived from the inner cell mass of the embryo, were first established in the mouse in 1981 and in humans in 1998 [329, 330]. Since then, protocols have been developed to efficiently differentiate ESCs into MSC-like progenitors capable of further differentiation into hyaline-like chondrocytes and fibrochondrocytes [319, 323–326]. In cartilage tissue engineering applications, human ESCs were differentiated within hydrogels and transplanted into osteochondral defect models in immunodeficient rats, where they contributed to the regeneration of hyaline-like cartilage [331, 332]. Human ESCs have also been used to generate scaffold-free, self-assembled neocartilage with fibrocartilage-like characteristics in vitro [323, 324, 333]. Furthermore, a scaffold-free pellet culture system using human ESCs produced hyaline-like neocartilage with a tensile Young's modulus of 0.91 MPa [334]. Notably, human ESC-derived chondrocytes seeded in collagen membranes successfully repaired full-thickness articular cartilage defects in a porcine model, resulting in hyaline-like tissue formation [335]. Despite these promising preclinical outcomes, no clinical trials using human ESCs for cartilage repair have been initiated to date. Because of ethical concerns associated with ESC use and the discovery of iPSCs, translational efforts have increasingly focused on iPSCs.

In 2006 and 2007, the generation of murine and human iPSCs from adult fibroblasts was reported, respectively [336, 337]. Since their discovery, iPSCs have been generated by reprogramming various somatic cells, such as fibroblasts and peripheral blood mononuclear cells, into a pluripotent state through the introduction of defined transcription factors [336, 337]. iPSCs circumvent the ethical concerns associated with ESCs and allow for the creation of patient-derived or donor-derived pluripotent cell lines suitable for autologous or allogeneic applications, respectively. Since their discovery, various methods have been developed to generate both iPSC-derived progenitors and neocartilage. The generation of neural crest-like progenitors has been demonstrated using both human iPSCs and ESCs [338, 339]. In addition, chemically defined media have been used to generate iPSC- and ESC-derived progenitors with paraxial mesoderm characteristics, which can subsequently be differentiated into hyaline-like cartilage [321, 338].

In tissue engineering applications, scaffold-free, hyaline-like cartilage particles derived from iPSCs were implanted into osteochondral defects in immunodeficient rodent and murine models and an immunosuppressed porcine model [320]. These iPSC-derived cartilage particles were also evaluated in chondral defects of the knee using cynomolgus monkey iPSCs in an allogeneic manner [340]. In a landmark first-in-human clinical trial, allogeneic human iPSC-derived cartilage particles were implanted into four patients with knee articular cartilage defects, demonstrating safety and clinical improvement at 12 months [341]. Although human iPSCs have not been used to generate elastic cartilage of the outer ear, a notable study succeeded in forming inner ear organoids from human iPSCs containing cartilaginous tissue of neural crest origin [342]. For tracheal cartilage tissue engineering, human iPSCs have been combined with a collagen sponge and polypropylene scaffold to form tracheal implants [343]. Other studies have utilized electrospun polycaprolactone nanofibers and 3D-printed polycaprolactone microfibers in combination with iPSC-derived progenitors and chondrocytes to engineer the tracheal mucosa and cartilage [344]. As of 2024, a review identified 110 clinical trials involving human iPSCs, but only two targeting cartilage repair: one involved iPSC-derived progenitors for intra-articular injections (non-tissue engineering), and just one involved the aforementioned tissue engineering product using iPSC-derived cartilage particles [345].

In summary, ESCs and iPSCs have been explored as alternative cell sources for cartilage tissue engineering, with one phase I clinical trial completed to date in Japan. Several barriers continue to limit broader clinical adoption. These include the risk of residual pluripotent cells leading to tumor formation, variability in differentiation efficiency, batch-to-batch inconsistency, and the costs associated with large-scale manufacturing under GMP [346]. Notably, in Japan, a national initiative (iPSC stock project) has focused on generating GMP-grade, HLA-homozygous iPSC lines that match the Japanese population at major HLA loci [347]. These human iPSCs have already been utilized in at least 14 clinical trials, including the one for cartilage [348]. These efforts, including the manufacturing of GMP-grade cell lines, reflect a growing infrastructure to support future clinical applications. Continued progress in addressing translational challenges will shape the future impact of ESC- and iPSC-based cartilage tissue engineering.

3.1.4 Cartilage-Derived Progenitor Cells (CPCs)

CPCs (also referred to as chondroprogenitor cells, cartilage progenitor cells, or cartilage stem cells) represent a population of clonogenic, multipotent, and self-renewing cells residing within or adjacent to cartilaginous tissues. These cells have been identified within hyaline cartilage, fibrocartilage, and elastic cartilage and are thought to contribute to tissue development, maintenance, and repair in normal development and life. CPCs are generally characterized by their high proliferative capacity and a propensity to differentiate toward the chondrogenic lineage [349]. CPCs have been isolated and purified through primarily three techniques: fibronectin adhesion assay (FAA), migratory assay (i.e., explant culture), and cell sorting.

FAA has been used to isolate cell populations with high affinity for fibronectin, and these cells exhibit strong colony-forming ability and can be highly expanded while maintaining multipotency [350–352]. In tissue engineering, FAA-derived CPCs have been tested in large-animal models such as caprine and equine models with encouraging results [351, 353, 354]. For example, FAA-derived CPCs were compared to the mixture of articular chondrocytes from which they were derived in a caprine model. Here, full-thickness articular cartilage was digested to yield cells from all zones of cartilage. Then, these cells were either expanded and seeded to a collagen membrane or expanded and seeded after purification through FAA. Although FAA-derived CPCs underwent a higher degree of cell expansion than the bulk articular chondrocytes before seeding, the cartilage repair observed from both groups was similar after 20 months [351]. In another study using an equine model, both autologous and allogeneic FAA-derived CPCs were delivered in fibrin to treat chondral defects [353]. After 12 months, autologous CPCs formed superior cartilage repair tissue compared to their allogeneic counterparts. FAA-derived CPCs have also been characterized from both the inner and outer zones of the knee meniscus [355]. In an ex vivo human model, FAA-derived CPCs from articular cartilage have been used to bridge meniscal tears [356]. Thus, FAA-derived CPCs show promise as a potentially abundant cell source for cartilage and meniscus repair.

The migratory assay, an explant culture technique, has been used to isolate proliferative cells with potential for high chondrogenicity. Migratory CPCs are obtained by plating cartilage fragments, often predigested with collagenase, onto culture plates and collecting the outgrowth of migrating cells. These cells exhibit multipotency and may possess greater chondrogenic potential than FAA-derived CPCs [357, 358]. Preclinical studies using migratory CPCs have largely been limited to small animal models. In one study, human migratory CPCs were combined with PRP and activated with thrombin to form constructs that were implanted into leporine cartilage defects, resulting in superior cartilage repair compared to PRP mixed with BM-MSCs or expanded articular chondrocytes [359, 360]. Another study demonstrated that migratory CPCs from human articular cartilage, expanded with platelet lysate and implanted as pellets or on polyglycolic acid and hyaluronan scaffolds in an immunodeficient murine model, induced hyaline-like cartilage formation [361]. A separate study comparing migratory CPCs derived from fetal and adult cartilage found that fetal-derived CPCs exhibited superior chondrogenic and proliferative capacities [362]. Migratory CPCs are also promising because they have been isolated from other cartilage tissues, such as the annulus fibrosus and knee meniscus [363].

The third method to isolate CPCs is using cell sorting techniques such as fluorescence-activated cell sorting (FACS). For example, CD146, a cell surface glycoprotein also known as melanoma cell adhesion molecule, has been used to isolate CPCs from human articular cartilage, with the sorted CD146 + cells demonstrating higher chondrogenic capacity than AD-MSCs and unsorted articular chondrocytes [364]. A small subpopulation of CPCs was isolated through cell sorting from the superficial zone of articular cartilage, exhibiting a strong propensity to express a superficial zone articular chondrocyte phenotype upon differentiation [365]. For elastic cartilage repair, CPCs have been obtained from the human auricular perichondrium by clonally expanding cells positive for both CD44 and CD90 [7]. These cells formed elastic cartilage that was stable after extended in vitro expansion. Following in vivo transplantation into immunodeficient mice, mature elastic cartilage resembling native auricular cartilage was observed after 10 months, complete with a surrounding perichondrium layer. Cell sorting has enabled the isolation of CPCs with enhanced chondrogenic potential for both hyaline and elastic cartilage applications.

Despite growing knowledge and the identification of CPCs across multiple tissue types, significant challenges remain for their clinical translation. Isolation methods differ widely among studies, and further research is needed to determine optimal tissue sources and expansion protocols capable of yielding clinically relevant cell numbers. Nonetheless, CPCs residing in or adjacent to cartilaginous tissues represent a promising alternative to

MSCs derived from non-cartilaginous sources because CPCs have a strong commitment to the chondrogenic lineage.

3.2 Scaffold-Based Approaches

Scaffold-based approaches to cartilage tissue engineering utilize three-dimensional (3D) biomaterials to provide structural support, guide tissue organization, and replicate key features of the native ECM. These scaffolds facilitate cell attachment, proliferation, differentiation or redifferentiation, and tissue maturation. They are especially important in engineering cartilages with defined geometries, such as the auricular and nasal cartilages, and those requiring immediate support of load-bearing after implantation. Scaffolds may be fabricated from natural materials, synthetic polymers, or composites that combine the advantages of both. Innovations in scaffold fabrication, including electrospinning, 3D bioprinting, and 3D weaving, have improved the ability to reproduce the complex spatial architecture of native cartilages [231]. These technologies enable zonal and anisotropic organization and accommodate cartilage-specific and patient-specific geometries. This section examines scaffold materials commonly used in cartilage tissue engineering, emphasizing their application to different cartilage subtypes and their progress toward clinical translation.

3.2.1 Natural Scaffolds

Commonly used naturally derived biomaterials include alginate, agarose, collagen, chitosan, silk fibroin, fibrin, and HA. This section briefly describes each of these materials and highlights their applications in engineering hyaline cartilage, fibrocartilage, and elastic cartilage.

Alginate is a polysaccharide extracted from algae. It can encapsulate cells in 3D, preserving a rounded cell morphology, and support the redifferentiation of cells expanded in monolayer. Alginate encapsulation has been used to redifferentiate/differentiate various cell types, including passaged chondrocytes, MSCs, and CPCs [366, 367]. Alginate is biocompatible, and filtered medical-grade formulations are available for sterile applications [368]. In a leporine study, full-thickness articular cartilage defects treated with alginate gel containing allogeneic chondrocytes resulted in hyaline-like cartilage repair at 6 months, whereas defects treated with alginate alone produced only fibrous tissue [369]. While alginate may degrade rapidly in vivo and potentially hinder new tissue formation, recent crosslinking and 3D bioprinting strategies attempt to improve stability and tissue formation. A double-crosslinked alginate hydrogel was developed for use in 3D bioprinting with human UC-MSCs, demonstrating neocartilage formation at 4 weeks [368]. Alginate is an attractive biomaterial for cell encapsulation and 3D bioprinting.

Agarose is a seaweed-derived polysaccharide hydrogel. It has a temperature-sensitive solubility in water. Similar to alginate, it can maintain a rounded chondrocyte shape in 3D culture. Agarose has been extensively used in vitro to study chondrocyte mechanobiology because its continuous gel matrix transmits mechanical loads to embedded cells [370, 371]. For example, dynamic compression of chondrocyte-seeded agarose in vitro increases production of GAGs and collagen [372]. An alginate-agarose hydrogel used in ACI has been investigated in phase III clinical trials in Belgium and France [373, 374]. While the 2-year outcomes of the Belgian trial indicated significant clinical improvement, the corresponding study in France reported significantly worse functional outcomes compared to mosaicplasty. These mixed clinical findings highlight the need for further evaluation of agarose-based hydrogels but support their potential for cartilage tissue engineering.

Collagen is a major component of the ECM in connective tissues, including the various cartilages, that can be isolated and used as scaffolds. Type I collagen, typically derived from tendons or skin, is widely used to fabricate sponges or membranes for cartilage repair and must be processed to reduce immunogenicity [375]. ACI products often employ collagen scaffolds. For example, the MACI product uses autologous chondrocytes seeded on a type I/III collagen membrane (porcine collagen). One study reported that articular chondrocytes on type I collagen scaffolds, rather than type II collagen scaffolds, produced more GAGs and expressed higher levels of chondrogenic gene expression [376]. Collagen scaffolds have also been explored for fibrocartilage tissue engineering. An acellular CMI is used in tissue engineering with the addition of cells. Research in sheep and clinical work in humans showed that seeding autologous BM-MSCs on CMI can enhance the repair of meniscus tears in avascular regions compared to an unseeded CMI [299]. In auricular cartilage tissue engineering, collagen gels or sponges seeded with chondrocytes have been studied [377]. For example, a patient-specific ear implant was generated from a 1:1 mixture of human auricular chondrocytes and MSCs embedded in a type I collagen hydrogel, using 3D printed ear molds [378]. After 3 months of testing in immunodeficient rodent models, the ear shape was maintained with a layer of perichondrium, proteoglycan deposition, and an elastin fiber network. In another study, decellularized auricular cartilage scaffolds containing type II collagen (AuriScaff) were made with the selective depletion of elastin fibers. These scaffolds have been used with human articular chondrocytes and AD-MSCs, but for the purpose of hyaline cartilage repair. Collagen is widely used in scaffolds because of its biocompatibility and its native presence in the ECM.

Chitosan is a polysaccharide derived from chitin (crustacean shells) via deacetylation. It is structurally similar to GAGs and is biocompatible and biodegradable, with a degradation rate tunable by its degree of deacetylation. Chitosan is often formulated as an injectable thermogelling solution that solidifies at body temperature, making it attractive for use in minimally invasive procedures [379]. Chitosan hydrogels seeded with articular chondrocytes have shown promise in vivo cartilage repair. For example, a study implanted chitosan hydrogel, either alone or seeded with cells, into full-thickness knee cartilage defects in an ovine model [379]. By 24 weeks, both treatments showed repair not observed in untreated sites, with the cell-seeded group demonstrating superior outcomes. Chitosan-based carriers have also been tested for delivering cells to the IVD and other fibrocartilages. In an ovine model, a chitosan hydrogel with AD-MSCs was injected into damaged IVDs, leading to stabilization and decreased degeneration compared to untreated controls at 12 months [380]. For meniscus repair, pure chitosan is less commonly used, but composite scaffolds (e.g., chitosan combined with gelatin or collagen) have been explored [381, 382]. Overall, chitosan's gelation ability, biodegradability, and biocompatibility make it a suitable material for cartilage tissue engineering.

Silk fibroin (most commonly from silkworm cocoons or spiders) is a fibrous protein polymer noted for its high tensile strength, slow degradation, and biocompatibility. Silk can be processed into porous sponges, fibers, or hydrogels. Silk fibroin is also suitable for 3D bioprinting [383]. Studies have shown that silk fibroin scaffolds can support chondrogenesis. For example, they have been used with human articular chondrocytes and MSCs to demonstrate attachment, proliferation, and differentiation within the scaffolds [384]. Silk fibroin-based bioinks have also been used to generate complex organ structures, including a ring-like cartilaginous trachea with a hyaline-like phenotype [385]. In meniscus tissue engineering, knitted or electrospun silk scaffolds have been studied because their toughness helps to reconstruct the load-bearing collagen network of the knee meniscus [386]. Silk fibroin-gelatin-based bioink has been used to 3D print meniscus scaffolds seeded with fibrochondrocytes from the knee meniscus [387]. 3D bioprinting has been used to fabricate biomimetic annulus fibrosus from silk fibroin-based scaffold seeded with primary fibrochondrocytes from the annulus fibrosus and AD-MSCs [388, 389]. For elastic cartilage, silk has been used to fabricate ear-shaped frameworks. For example, a silk-gelatin-blended ink has been used with auricular chondrocytes to 3D bioprint ear-shaped specific neocartilage [390]. An alginate and silk polymer scaffold seeded with auricular chondrocytes yielded cartilaginous tissue with the shape of a human auricle in a leporine model with good structural fidelity [391]. In summary, silk fibroin provides a strong, slowly-degrading scaffold that can be used to form neocartilage.

Fibrin is a natural protein polymer formed from fibrinogen through thrombin-mediated crosslinking (like in a blood clot). Furthermore, fibrin is FDA-approved for use as a hemostat, sealant, and adhesive, and is commercially available [392]. In orthopaedic applications, fibrin sealants are frequently used, including their use in securing ACI products into cartilage defects [393]. Fibrin's properties of rapid gelation, non-toxic degradation, and biocompatibility make fibrin-based hydrogels appealing for use in cartilage tissue engineering [394]. However, they also shrink and degrade rapidly in vitro and in vivo and have poor mechanical properties. Nevertheless, fibrin sealants are used as scaffolds to deliver cells in tissue engineering. Chondron (Sewon Cellontech), an ACI product with cultured autologous chondrocytes mixed in fibrin sealant (fibrin ACI), is cleared for marketing in South Korea and India [231, 395, 396]. More recent advances have focused on incorporating fibrin with other naturally-derived biomaterials, such as alginate and collagen, or with synthetic polymers [394]. In summary, fibrin is valued for its biocompatibility and extensive clinical use, although its mechanical properties often require enhancement through the incorporation of other biomaterials.

HA is a non-sulfated GAG present in native cartilages. HA-based hydrogels are frequently used in cartilage tissue engineering and can be chemically modified to confer diverse mechanical and biological properties [397]. Clinically, Cartistem, which consists of allogeneic UC-MSCs in HA hydrogel, is approved for marketing in South Korea [314]. HA-tyramine conjugates have been used with BM-MSCs to show that increasing the degree of cross-linking and matrix stiffness shifts MSC differentiation from a hyaline cartilage phenotype toward a more fibrocartilaginous phenotype [398]. In another study, bovine articular chondrocytes embedded in HA hydrogel crosslinked with an elastin-like protein exhibited enhanced GAG deposition at higher HA concentrations, although elastin content was not evaluated [399]. Another study, using auricular chondrocytes in HA hydrogel, demonstrated strong type II collagen staining 12 weeks after implantation into an immunodeficient murine model [400]. Although HA-based hydrogels are used in clinical applications, their use in fibrocartilage and elastic cartilage engineering remains underexplored and presents opportunities for further development.

3.2.2 Synthetic Scaffolds

Synthetic scaffolds offer tunability in mechanical properties, degradation rates, and biological activity. Polymers such as poly-glycolic acid, poly-lactic acid, poly-caprolactone, poly-urethane, and poly-ethylene glycol have been the most commonly used in cartilage tissue engineering. These polymers can be formed into porous scaffolds, non-woven meshes, or felts, allowing scaffold shape and architecture to be extensively manipulated. Poly-glycolic acid is an aliphatic polyester that degrades by hydrolytic scission. Total degradation can occur within 4–12 months, with degradation products being naturally resorbed into the body [401, 402]. Poly-glycolic acid scaffolds themselves are generally weak, but cells seeded onto the scaffolds can secrete ECM to provide mechanical integrity. Poly-lactic acid is another aliphatic polyester that generally degrades more slowly than poly-glycolic acid, over a period of 12 months to more than 2 years [403]. The degradation products of poly-lactic acid are resorbable and have been utilized as materials for sutures. Poly-lactic acid scaffolds are primarily fabricated as non-woven meshes. Poly-lactic-co-glycolic acid is a copolymer composed of poly-glycolic acid and poly-lactic acid monomers, and the ratio of each monomer affects material properties [404]. A formulation with higher poly-lactic acid content will degrade more slowly than one with higher poly-glycolic acid content. A 75/25 (poly-lactic acid/poly-glycolic acid) copolymer showed degradation within 4–5 months, whereas a 50/50 copolymer degraded in just 1–2 months [403]. For example, 50/50 Poly-lactic-co-glycolic acid was tested in a caprine model without cells, resulting in no filling of cartilage defects at 16 weeks [405]. Poly-caprolactone is another popular synthetic polymer for cartilage tissue engineering with longer degradation times than poly-glycolic acid, poly-lactic acid, or poly-lactic-co-glycolic acid, and is generally stronger [406]. Poly-caprolactone can be formed into threads for meshes or felts, or fabricated into porous scaffolds. Poly-caprolactone degrades through hydrolytic scission, but this process can take 1–2 years for complete degradation [403]. While several factors influence the selection of synthetic polymers, including mechanical properties, processability, biocompatibility, and support for cell attachment and proliferation, degradation time remains a key consideration and can be modulated by varying material combinations.

In addition to the commonly used aliphatic polyesters, other synthetic polymers such as poly-ethylene glycol and poly-urethane offer unique functionalities. Poly-ethylene glycol is a polymer often incorporated into copolymers for its high hydrophilicity, which prevents adsorption of proteins and cells. Poly-ethylene glycol can be used to limit cell attachment or to reduce adsorption of antibodies and other proteins, thereby lessening immune responses. Poly-ethylene glycol does not naturally degrade in the body and is thus typically used in copolymerization with other materials. Poly-urethane is a versatile synthetic polymer used in cartilage tissue engineering for its tunable mechanical properties and biocompatibility. Poly-urethane can be formulated to exhibit elastic or rigid properties and can be processed into porous scaffolds or hydrogels [407]. Biodegradable forms of poly-urethane are typically synthesized using aliphatic diisocyanates and degradable soft segments such as poly-caprolactone or poly-ethylene glycol, allowing for hydrolytic and enzymatic degradation over time. Its mechanical resilience and ability to support cell adhesion make poly-urethane a promising scaffold material, especially in load-bearing cartilage applications. Poly-ethylene glycol and poly-urethane are additional synthetic materials that can be used to form scaffolds with controlled immune compatibility, mechanical properties, and structural features.

Advanced fabrication methods such as electrospinning and 3D weaving have been used to create structurally organized scaffolds that better mimic the complexity of native cartilages. Electrospinning, a fiber production technique, has been employed to fabricate anisotropic scaffolds that replicate the internal organization of cartilage by layering horizontally, randomly, and vertically aligned poly-caprolactone fibers in a multilayered configuration [408]. This method has also enabled the integration of cartilage-derived matrix into poly-caprolactone scaffolds, which enhanced GAG and collagen production by AD-MSCs compared to poly-caprolactone-only controls [409]. 3D weaving techniques have been applied to develop osteochondral scaffolds by combining woven poly-caprolactone fibers with MSCs and bone, promoting distinct cartilage and bone phenotypes within a single construct [410]. These advanced fabrication techniques allow precise architectural control and spatial patterning of bioactivity, critical for tissue engineering.

3.2.3 Composite Scaffolds

Composite or hybrid scaffolds are formed from a combination of natural and synthetic materials. In many studies, a synthetic mesh is infused with natural materials containing cells. These scaffolds are particularly useful in recreating zonal architecture and functional gradients that mimic the structure and function of native cartilage. For example, a microscale 3D weaving technique was used to fabricate a 3D structure using poly-glycolic acid and infused with hydrogels (agarose or fibrin) containing articular chondrocytes [411]. These 3D composite scaffolds, without cells, showed anisotropic and viscoelastic properties similar to native articular cartilage and, with chondrocytes, the ability to maintain such chondrocytes in a rounded morphology with a chondrogenic phenotype. For tissue engineering of the knee meniscus, decellularized meniscus ECM and polycaprolactone have been used through electrospinning to make scaffolds for use with fibrochondrocytes obtained from the knee meniscus [412]. In another study, composites using 3:1 silk fibroin and polyvinyl alcohol for the knee meniscus have been used to create scaffolds that support fibrochondrocyte proliferation and ECM secretion [413]. For elastic cartilage engineering, AD-MSCs and auricular chondrocytes seeded to alginate hydrogel were infused into a poly-caprolactone scaffold [414]. Other composites, such as poly-lactic-co-glycolic acid mesh/collagen sponge, poly-lactic acid/collagen/chitosan, and ECM/poly-lactic-co-glycolic acid, have been explored as scaffolds suitable for seeding of chondrocytes and MSCs [415–417]. Composite scaffolds support cell attachment while providing biomechanical and degradation properties that match the functional needs of cartilage implants.

3.3 Scaffold-Free Approaches

Scaffold-free (scaffoldless) approaches to cartilage tissue engineering rely on the inherent ability of certain cells to organize and generate ECM in the absence of exogenous biomaterials [214]. These strategies often reflect early developmental processes, particularly mesenchymal condensation and chondrogenic differentiation, and promote tissue formation through direct cell–cell and cell–matrix interactions [418]. By avoiding the use of exogenous scaffolds, scaffold-free methods minimize the risks associated with immune response, inflammation, and material degradation, potentially leading to lower regulatory burden. Thus, scaffold-free approaches to cartilage tissue engineering have been explored extensively in both autologous and allogeneic applications.

Scaffold-free approaches can be categorized into two methods: self-organization and self-assembly [214]. Self-organization, by definition, requires the input of external energy or force toward tissue formation. Pellet culture, micromass culture, aggregate culture, and cell sheet technologies are examples of self-organization methods. Self-assembly occurs without the input of external energy and includes methods to form spheroids in U/V-shaped wells and the self-assembly process in agarose-coated wells. These methods have been applied across multiple cartilage types, including hyaline cartilage for joint repair, fibrocartilage for knee meniscus and TMJ disc repair, and elastic cartilage for auricular applications. The following sections examine the development and translational potential of scaffold-free platforms for cartilages.

3.3.1 Self-Organization Methods

Self-organization methods require an input of energy, such as through centrifugation, rotation, temperature change, or physical manipulation. Centrifugation is one method that allows the condensation of cells into high-density suspensions. This method mimics the mesenchymal condensation stage of cartilage development, enabling robust ECM production in a compact space. Traditionally, cell suspensions are transferred to conical tubes or U/V-shaped wells and centrifuged. This method has been widely used as a screening assay for measuring the ability of cells to form hyaline cartilage in vitro [419]. Clinically, CartiLife (Biosolution), containing autologous costal chondrocytes in pellet form, has been cleared for marketing since 2019 in South Korea with phase II clinical trials currently being conducted in the United States [231]. Articular chondrocytes have also been seeded to transwell inserts in plates, which can be centrifuged to form cartilage discs [420]. Aggregate culture using rotational culture is another form of self-organization. It has been applied as a method to redifferentiate monolayer expanded cells toward a chondrogenic phenotype [236, 421]. Self-organization yields neocartilage that can be used directly for implantation or to obtain cells with desired phenotypes for further application in cartilage tissue engineering.

Cell sheet technology has been applied to various tissues, including cartilage [422]. Cells are seeded onto temperature-responsive culture dishes coated with a thin layer of poly(N-isopropylacrylamide), a polymer that permits detachment of confluent cells as intact sheets upon temperature reduction [423]. Subsequent manipulation enables the stacking of these sheets to create thicker, implantable constructs. Autologous chondrocyte sheets have been fabricated using three layers of articular chondrocytes derived from the patient's own knee, whereas allogeneic sheets have been produced using a single layer of chondrocytes isolated from the supernumerary digits of juvenile polydactyly donors [246, 424]. First-in-human clinical studies in Japan have tested both autologous and allogeneic chondrocyte sheets for treating knee OA-associated cartilage defects, with no serious adverse events reported [243, 248]. While further trials are needed to confirm long-term safety and efficacy, the autologous approach has been approved under Japan's advanced medical care system, which allows conditional coverage by special insurance. Beyond chondrocytes, cell sheet technology has also been applied to other cell types such as BM-MSCs and iPSC-derived progenitors for cartilage repair [328, 425–427]. While hyaline cartilage regeneration has been the main goal, other fibrocartilage and elastic cartilages may be targeted with cell sheet technology because of its success thus far.

3.3.2 Self-Assembly Methods

Among self-assembly methods, the self-assembling process has been applied successfully in generating mechanically robust neocartilage [214, 236, 237]. The self-assembling process involves seeding cells at high density into molds made of nonadherent substrates such as agarose (Fig. 3.2) [214]. Through the minimization of free energy without the input of external forces, cells bind to each other through cell adhesion molecules, migrate, and produce ECM in phases resembling those in native tissue development. In cartilage tissue engineering, the self-assembling process has been applied to multiple cell sources, such as ESCs, MSCs, and chondrocytes, and across multiple species, including human, porcine, bovine, ovine, caprine, and leporine [236, 257, 258, 323, 428]. Notably, both human and minipig costal chondrocytes have been used to generate hyaline-like neocartilage at high passages with functional properties approaching native hyaline cartilage [234, 237, 261]. Allogeneic transplantation of hyaline-like neocartilage into TMJ disc perforations demonstrated defect closure in a minipig TMJ disc thinning model, demonstrating that hyaline-like neocartilage may be used to treat fibrocartilages [235]. The self-assembling process has been applied to fibrochondrocytes as well, where a 1:1 co-culture of articular chondrocytes and fibrochondrocytes from the knee meniscus has been used to form neocartilage with the meniscus's wedge shape and anisotropy [269, 271]. Neonatal porcine auricular chondrocytes have also been used to form self-assembled neocartilage with properties of elastic cartilage [8]. Human BM-MSCs have been used to form self-assembled cartilaginous rings toward the development of tracheal tissue [429]. In short, the self-assembling process is a versatile method that has shown promise in producing cartilages of various types.

Spheroids are also formed through self-assembly as opposed to pellets, which are formed with centrifugation. Spheroid formation can be achieved through seeding cells into nonadherent U-shaped microwells, where the diameter of microwells influences the size of spheroids formed. Clinically, Spherox (formerly Chondrosphere), formed from autologous articular chondrocytes, is cleared for marketing in the EU and other countries [231, 430]. Micrometer-scale spheroids can be made using 3D printed agarose molds and using porcine knee articular chondrocytes and fibrochondrocytes from the knee menisci [431]. Human CPCs obtained from the auricular perichondrium have been used to fabricate microscopic spheroids for the treatment of external ear abnormalities [432]. Human nasal and auricular chondrocytes and CPCs obtained through explant culture have been used to form spheroids for engineering head and neck cartilages [433]. Interestingly, auricular spheroids were larger than nasal spheroids, with both containing elastin. Spheroids derived from human nasal chondrocytes have also been tested in treating degenerative

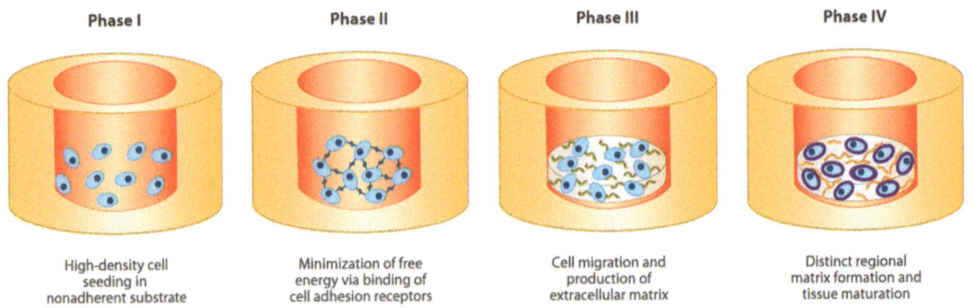

Fig. 3.2 Different phases in the development of the self-assembling process. In phase I, cells at high density are seeded in wells made from a nonadherent substrate. In phase II, cells bind to each other through the minimization of free energy and cell adhesion receptors. In phase III, cells migrate and produce ECM, and in phase IV, tissues mature through distinct regional matrix formation. Reproduced with permission [214]

IVD disease [434]. These findings demonstrate the broad applicability of spheroids for engineering the various cartilages.

3.4 Biochemical and Biomechanical Cues

The development of tissue-engineered cartilage products depends not only on the choice of cells and scaffolds but also on the application of external stimuli applied throughout tissue engineering (Fig. 3.3) [435]. This process involves determining the appropriate timing, duration, and quantity or magnitude of the stimuli. These stimuli comprise two main categories: biochemical and biomechanical cues. Biochemical cues, including growth factors, small molecules, and dissolved oxygen, can be manipulated to affect the different stages of the tissue-engineered cartilage process, i.e., expansion, differentiation or redifferentiation, and tissue maturation. Most biochemical factors used in cartilage tissue engineering are soluble, and they are covered here. Biomechanical cues, including mechanical forces, such as compression, tension, hydrostatic pressure, and shear, can be modulated to affect mechanical, structural, and organizational properties of neocartilage during tissue maturation. The following sections describe the applications of biochemical and biomechanical cues that have been studied in cartilage tissue engineering.

3.4.1 Common Biochemical Factors

Cartilage tissue engineering relies heavily on biochemical factors to direct proliferation, promote ECM production, and guide tissue development toward a functional phenotype [435]. Soluble biochemical factors are added to supplement the cell culture medium. For

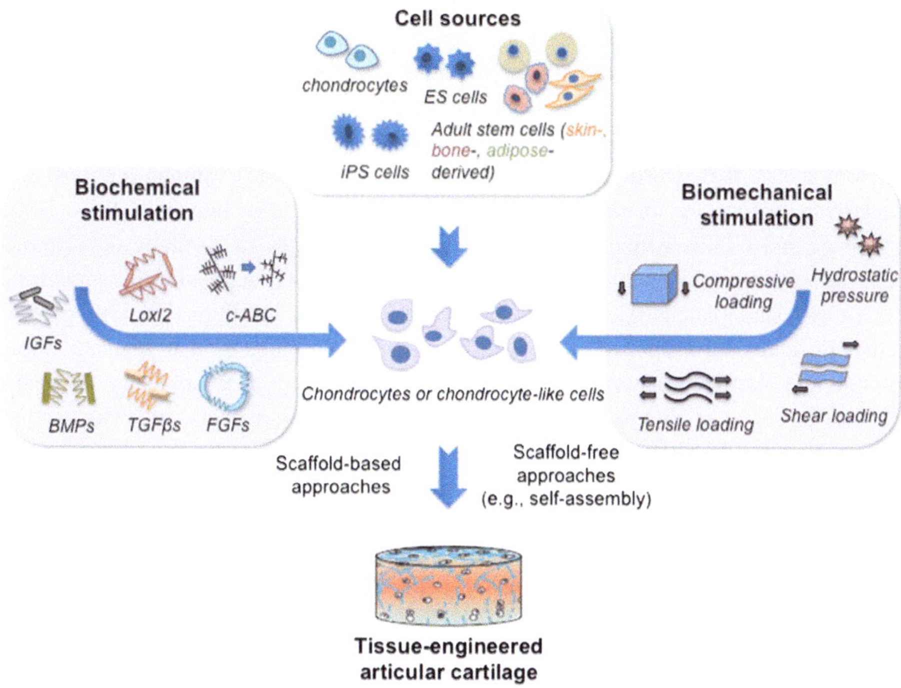

Fig. 3.3 Cartilage tissue engineering strategies. Cartilage tissue engineering typically entails the in vitro development of three-dimensional (3D) neocartilage using either chondrocytes or other cells that can be differentiated into chondrocyte-like cells. These cells are incorporated into scaffolds or self-organized/self-assembled using scaffold-free methods. Throughout the tissue engineering process, these cells may be exposed to biochemical and/or biomechanical stimuli to promote expansion, differentiation or redifferentiation, and tissue maturation. Numerous cell types, bioactive molecules, and physical stimuli have been investigated to support the tissue engineering of cartilages. Reproduced with permission [435]

example, serum, such as fetal bovine serum (FBS) or human serum, supports cellular function and ECM synthesis [436]. Serum provides a complex mixture of growth factors, hormones, and proteins; however, because of its variability and often xenogeneic origin, defined supplements that partially or fully replace serum are often preferred [436]. Ascorbic acid is essential for collagen production, acting as a cofactor for enzymes involved in collagen crosslinking [437]. Insulin-transferrin-selenium (ITS) supports cell growth and reduces dedifferentiation [438]. L-proline is a key amino acid precursor for collagen biosynthesis and constitutes a third of the amino acid composition in collagen [439]. Dexamethasone, a synthetic glucocorticoid, enhances chondrogenic differentiation and promotes ECM production [440]. Glucose serves as the primary energy source for cells in culture, and its availability affects ECM synthesis, including GAGs [441]. Although not exhaustive, these supplements are among the most commonly used in cartilage tissue engineering. It remains important to carefully select and optimize the type and concentration of each supplement based on the specific cell type, the presence or absence of scaffolds, and the intended mechanical, structural, or organizational outcomes.

Beyond the supplements to basic media formulations, other biochemical cues such as growth factors, small molecules, and dissolved oxygen may be used to influence the different stages of the cartilage tissue engineering process [435]. Among the most studied growth factors are members of the TGF-β superfamily, including TGF-βs, BMPs, activin-A, and GDFs, which play central roles in all stages of cartilage development. Other important groups include fibroblast growth factors (FGFs), insulin-like growth factors (IGFs), and platelet-derived growth factors (PDGFs). Additionally, Wnt proteins, follistatin, and neurotrophin-4 act as developmental morphogens that regulate lineage commitment, cartilage patterning, and maturation. Low oxygen tension may be used to decrease the level of dissolved oxygen. Together, these components, in various combinations, help guide cells toward the desired phenotype required to tissue engineer the distinct forms of cartilage. Below, the use of biochemical cues is described as organized by different steps of cartilage tissue engineering: cell expansion, cell differentiation or redifferentiation, and tissue maturation.

3.4.2 Expansion

A critical step in cartilage tissue engineering is the expansion of cells. This can be achieved using traditional monolayer culture or specialized systems such as multi-layered flasks and even 3D bioreactors. Given the large number of cells required to generate neocartilage of relevant size, it is essential to expand cells while preserving their functionality for subsequent tissue engineering steps [236]. It is well established that chondrocytes rapidly dedifferentiate in monolayer culture, deviating from their native phenotype [442, 443]. Thus, cells must be expanded through protocols that either maintain their chondrogenic phenotype or maintain their ability to redifferentiate back to a chondrogenic phenotype. Thus, optimizing cell expansion requires a careful balance between maintaining cellular functionality and achieving efficient cell proliferation. ESCs, iPSCs, MSCs, and CPCs, which possess self-renewal capabilities of various degrees, are expanded under conditions that maintain their pluripotency or multipotency through the use of biochemical and other regulatory factors; these methods are described in detail elsewhere [336, 444–449].

The use of biochemical factors to expand chondrocytes has been extensively studied [442, 443]. For example, supplementation with FGF-2 during the expansion of articular chondrocytes resulted in up to a 2000-fold increase in cell yield while preserving their ability to redifferentiate post-expansion [450]. Further studies demonstrated that a combination of growth factors, TGF-β1, FGF-2, and PDGF-BB, applied during expansion of human articular chondrocytes isolated from OA cartilage, increased proliferation rates by 5.5-fold compared to cultures without growth factors [451]. In both human knee articular chondrocytes and costal chondrocytes, the same combination of TGF-β1, FGF-2, and PDGF-BB, used in media with 2% FBS during monolayer expansion, enabled expansion factors of 12.6 and 36.2 million, respectively, while maintaining the cells' ability to form hyaline-like neocartilage in vitro [236, 237]. In another study, human nasal chondrocytes were expanded in either serum-free media or 2% FBS supplemented with various combinations of TGF-β1, FGF-2, BMP-2, and IGF-1 [452]. The combination of 2% FBS, TGF-β1, and FGF-2 enhanced proliferation relative to serum-free conditions, while BMP-2 and IGF-1, although less effective for proliferation, were most successful in preserving chondrocyte morphology in monolayer. For fibrochondrocytes isolated from the knee meniscus, expansion in FGF-2 supplemented media improved the subsequent expression of type II collagen during aggregate culture [453]. These findings highlight the importance of biochemical factors in cellular expansion and underscore the need to explore expansion limits that yield clinically relevant cell numbers while preserving functionality [237].

3.4.3 Differentiation

Chondrogenic differentiation is the process by which stem cells or progenitor cells are directed to acquire a cartilage-specific phenotype appropriate for the target cartilage type. For ESCs and iPSCs, this process typically follows multiple stages, both in monolayer and in 3D culture, that recapitulate key aspects of embryonic development. In contrast, MSCs and CPCs are most commonly differentiated in 3D culture systems, such as pellet cultures or scaffold-based platforms like hydrogels. These methods utilize similar differentiation protocols as those used for chondrocyte redifferentiation, as discussed in Sect. 3.4.4. The use of specific biochemical factors has been extensively investigated to promote differentiation toward a hyaline-like cartilage phenotype, while only a limited number of studies have explored protocols aimed at directing differentiation toward fibrocartilage and elastic cartilage phenotypes.

Protocols to differentiate human ESCs into chondrocytes have been developed based on recapitulating key signaling pathways active during embryonic development, with progressive refinement of biochemical factors, their timing, and concentrations [319, 323, 324, 454, 455]. For example, a 14-day protocol for human ESCs involves the timed application of activin-A, Wnt3a, FGF-2, and BMP-4 in monolayer culture to induce mesendodermal differentiation, followed by BMP-4, FGF-2, follistatin, and neurotrophin-4 to promote mesodermal lineage commitment [319]. Nodules are then lifted into 3D, and BMP-4 is gradually substituted by GDF-5 to direct differentiation toward a chondrogenic phenotype. Many alternative protocols have employed various combinations of growth factors and small molecules, resulting in chondrocyte populations of varying purity and homogeneity

[320, 327]. Differentiation of ESCs must be tightly controlled to remove pluripotent cells or off-target cells that may form tumors or lack functionality.

Differentiation protocols for iPSCs have evolved from protocols for ESCs. For example, a step-wise protocol that follows the mesendodermal and mesodermal phenotypes has been shown to differentiate iPSCs to form scaffold-free, hyaline-like cartilage particles [320]. Wnt3a and activin-A were applied for 3 days to differentiate monolayer cultured iPSCs toward a mesendodermal phenotype, followed by the application of FGF-2, BMP-2, TGF-β1, and GDF-5 for 10 days. Under this 2D culture regimen, the cells spontaneously formed 3D nodules. Subsequently, cell nodules were detached and further cultured in suspension with BMP-2, TGF-β1, and GDF-5. After preclinical testing in rodent, porcine, and simian models, such iPSC-derived particles were tested in first-in-human trials in four patients with no serious adverse events reported [320, 340]. Another study demonstrated that as few as two small molecules may replace the various combinations of growth factors to direct chondrogenic differentiation from iPSCs to hyaline-like chondrocytes [456]. To generate annulus fibrosus-like fibrochondrocytes from iPSCs, growth factors such as TGF-β3, PDGF-BB, connective tissue growth factor, and IGF-1 have been tested [457]. Among these, the combination of TGF-β3 and PDGF-BB led to increased deposition of type I collagen and, interestingly, elastin, typically found in the outer annulus fibrosis [457]. While iPSC-derived cartilage particles have been explored for auricular reconstruction, no protocols have yet been reported that specifically drive iPSC-derived chondrocytes toward an elastin-expressing, elastic cartilage phenotype [457, 458]. Overall, the development of chondrogenic differentiation protocols for ESCs and iPSCs has advanced considerably, particularly for generating hyaline-like cartilage. Further development of protocols is needed to enable reliable differentiation toward fibrocartilage and elastic cartilage phenotypes.

In the chondrogenic differentiation of MSCs and CPCs, biochemical cues play a central role. Among these, TGF-β isoforms, TGF-β1, TGF-β2, and TGF-β3, are most commonly used to induce differentiation [459–463]. These factors are typically applied within 3D culture systems, such as in scaffolds or pellet cultures, which enhance cellular responses to chondrogenic signals. In human BM-MSCs, TGF-β3 was shown to be essential for chondrogenesis across various protocols, while late-stage (at 3 weeks) application of FGF-2 and parathyroid hormone-related protein (PTHrP) inhibited a hypertrophic phenotype [464]. In another study using human BM-MSCs, BMP-2, -4, or -6 with TGF-β3 and dexamethasone improved chondrogenic differentiation, while BMP-2 was most effective [465]. Other studies have looked at the transient application of TGF-β3, where 2- or 3-week applications of TGF-β3 were superior to continuous application during a 4-week culture period [466, 467]. The differentiation protocols for MSCs and CPCs need to be optimized to promote differentiation and inhibit hypertrophy in the case of hyaline cartilage.

3.4.4 Redifferentiation

Chondrogenic redifferentiation refers to the process of restoring the native phenotype of chondrocytes isolated from cartilage tissue. As with differentiation of MSCs and CPCs, TGF-βs are used alone or combined with other growth factors. For example, a combination of TGF-β1, BMP-2, and GDF-5 was found to induce redifferentiation of passaged human articular chondrocytes more effectively than any single factor or other combinations [468]. In other studies, culturing articular chondrocytes under reduced oxygen tension (2–5%) further enhanced redifferentiation [469, 470]. Auricular chondrocytes isolated from microtia patients and expanded to passage 5 were successfully redifferentiated in poly-glycolic acid/poly-lactic acid scaffolds using TGF-β1 and IGF-1, although the elastin expression was not assessed [471]. Redifferentiation of chondrocytes depends on the extent to which chondrocytes retain their ability to express a chondrogenic phenotype and is influenced by protocols for expansion as well as cell source and donor characteristics.

3.4.5 Neocartilage Maturation

Neocartilage maturation refers to the progressive development of neocartilage properties over time with increases in functional properties such as GAG and collagen contents and compressive and tensile properties. Biochemical treatment may also affect properties such as collagen crosslinking [472]. The degree of in vitro maturation of neocartilage typically varies with the duration of culture, and the extent of in vitro maturity affects in vivo results [473]. In addition to the factor of time, below are a few biochemical factors that have been shown to enhance functional properties of neocartilage.

Application of biochemical factors, such as chondroitinase-ABC (c-ABC), lysyl oxidase-like 2 (LOXL2), and agarase, has been investigated. Counterintuitively, c-ABC depletes GAG content at the time of application but enhances collagen synthesis and improves tensile properties during subsequent tissue maturation. For example, application of c-ABC to self-assembled neocartilage at both 2 and 4 weeks during a 6-week maturation period resulted in 78% and 64% increases in tensile modulus and UTS, respectively, over untreated controls [472]. Similarly, c-ABC treatment had an additive effect to the treatment of TGF-β1 on tensile properties and a synergistic effect on collagen content [474]. The addition of LOXL2, which promotes collagen crosslinking, further enhanced the system and led to 245% and 185% increases in tensile modulus and UTS, respectively, compared to untreated controls [475]. Enzymatic removal of scaffolds during maturation has also been shown to improve mechanical properties. Agarose scaffolds seeded with bovine articular chondrocytes were applied agarase at day 42 and were cultured to day 91, resulting in ~25% more DNA, ~60% more collagen, and ~40% higher dynamic compressive modulus compared to untreated controls [476]. While a few soluble biochemical factors have been identified that substantially enhance functional properties, their effects may be combined with biomechanical stimuli, as discussed in the following sections.

3.4.6 Common Biomechanical Factors

Biomechanical cues may offer additional ways to enhance functionality of neocartilage, typically complementing biochemical cues. As with biochemical cues, biomechanical cues must be optimized based on the cell type and the scaffold, if used. Optimization may involve the modulation of various factors, such as magnitude, duration, frequency, and timing [477]. Biomechanical cues include direct compression, tension, hydrostatic pressure, shear, and others [477]. These methods have been shown to enhance mechanical, structural, and cellular properties of neocartilage. The sections below highlight the most commonly applied biomechanical factors in cartilage tissue engineering.

3.4.7 Direct Compression

Direct compression mimics the mechanical force that joint articular cartilages experience under load (Fig. 3.4). This can be applied using simple weights or mechanical systems that deliver either passive loading (a constant force) or dynamic loading (a force that varies with time). Compressive forces have been found to increase collagen and GAG contents and improve compressive properties of the stimulated neocartilage. In cell types with a propensity toward hypertrophy, compression has been shown to inhibit hypertrophy.

Studies have typically employed direct compression at ~10% strain, although as high as 50% strain has been examined, or report stress as a measurement of compressive load. Dynamic compression of 5–20% strain on leporine AD-MSCs encapsulated in alginate hydrogel showed that 10% strain was optimal for chondrogenic differentiation [478]. Similarly, dynamic compression at 10% strain applied to leporine BM-MSCs in agarose induced chondrogenic differentiation as effectively as TGF-β1 treatment alone [479]. In human BM-MSCs in HA hydrogels, dynamic compression at 10% strain improved compressive properties as well as reduced the expression of hypertrophic markers [480]. Dynamic compression at 10% strain applied to human articular chondrocytes seeded to type I collagen hydrogels improved "elastic stiffness" compared to non-loaded controls [481]. In a study exploring the effect of dynamic compression in alginate scaffolds seeded with chondrocytes from either the superficial zone or middle/deep zone of articular cartilage, 50% strain increased lubricin, GAG, and compressive properties for neocartilage using superficial zone chondrocytes but not using middle/deep zone chondrocytes [482]. In a study using human iPSC-derived progenitors encapsulated in poly-ethylene glycol-chondroitin sulfate hydrogels, dynamic compression at 5% strain with BMP-2 and/or TGF-β3 promoted a stable chondrogenic phenotype with downregulation of type X

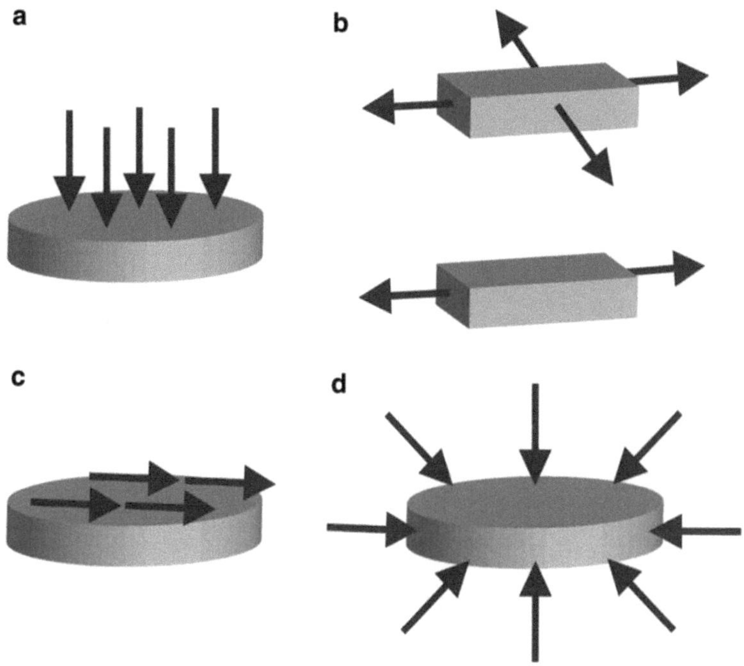

Fig. 3.4 Common types of biomechanical factors that may be applied to enhance neocartilage properties. **a** Direct compression, **b** biaxial tension (top) and uniaxial tension (bottom), **c** shear, and **d** hydrostatic pressure. Reproduced with permission [477]

collagen [483]. Dynamic compression at 10 kPa applied to human SY-MSCs encapsulated in agarose hydrogels showed that loading at day 1 downregulated the expression of chondrogenic markers while loading at day 21 enhanced such markers and suppressed hypertrophic markers [484]. In another study, passive compression applied at 3.3 or 5.0 kPa significantly increased self-assembled neocartilage properties by 42% and 48%, respectively, compared to controls, but no improvement was observed at 6.7 kPa [258]. These studies indicate that optimal strain/stress applied should be tailored to the cell type and source (e.g., anatomical location), scaffold use, and timing of application.

For fibrocartilage tissue engineering, dynamic compression has been shown to increase GAG and collagen contents as well as mechanical properties. Dynamic compression at 7–15% strain was applied to tissue-engineered menisci fabricated from bovine fibrochondrocytes seeded in alginate scaffolds. Dynamic compression applied for 2 weeks increased GAG content by 2.4–3.2 fold and compressive modulus by 1.8–2.5 fold, while after 6 weeks of total loading, GAG content and compressive modulus decreased while collagen content increased [485]. In a follow-up study, 15% strain was applied for 2 weeks followed by 4 weeks of static culture, yielding improvements in collagen bundle formation, 2.8-fold increase in GAG content, twofold increase in collagen content, and

4.3-fold increase in compressive modulus [486]. Further improvements using condyle-shaped platens for dynamic compression at 10% strain and anchoring (passive tension) of tissue-engineered menisci at the ends resulted in significant improvement in GAG content by 200–250%, collagen content by 40–55%, equilibrium modulus by 1000–1800%, and tensile moduli in radial and circumferential directions by 500 and 1200%, respectively, demonstrating anisotropy [487, 488]. Similarly, 10% strain applied in conjunction with biochemical factors to self-assembled neocartilage fabricated from a 1:1 mixture of articular chondrocytes and fibrochondrocytes from the knee meniscus resulted in increases of collagen content by fourfold, compressive properties by 3 to fourfold, and tensile properties of 4 to sixfold [489]. Dynamic compression, applied at the right timing and magnitude, may be an effective way to improve mechanical properties of engineered hyaline cartilage and fibrocartilage.

For elastic cartilage tissue engineering, only a few studies have explored the application of mechanical stimuli, including compression [192]. Dynamic compression at 10% strain showed that auricular chondrocytes in HA hydrogels exhibited increases in gene expression for type II collagen and aggrecan compared to unstimulated controls, but these increases were smaller compared to similarly treated articular chondrocytes [490]. Because auricular cartilage is non-load bearing in typical conditions, these results indicate that inherent differences in cell sources may affect the way neocartilage responds to mechanical stimuli. Overall, application of dynamic compression must be tailored to the tissue engineering system including cell source, scaffold, other stimuli used, and target cartilage type.

3.4.8 Tension

Tension refers to the forces applied when tissues are pulled outward (Fig. 3.4). Tensile forces can be applied uniaxially (in one direction) or biaxially (in two directions) using flexible cell culture surfaces or mechanical stretching devices that deform tissues. When delivered under appropriate loading conditions, tension enhances neocartilage formation by increasing ECM production, increasing collagen fiber alignment and thickness, and improving tensile strength.

Large increases in tensile mechanical properties as well as anisotropy can be obtained through the application of uniaxial tension. For example, application of uniaxial continuous tension at 12–15% strain during tissue maturation improved functional properties for both bovine and human self-assembled, hyaline-like neocartilage from articular chondrocytes [491]. With bovine articular chondrocytes, application of continuous tension improved both tensile Young's modulus and UTS by close to six times the control values, reaching 6.4 MPa and 2.3 MPa, respectively, and resembling native tissue values. The neocartilages were also anisotropic with tensile properties significantly different in the directions parallel and perpendicular to the uniaxial tension applied. For human neocartilage, Young's modulus and UTS reached 2.3 MPa and 1.2 MPa, respectively, with similar anisotropic properties. The significant improvement in tensile properties through the application of tension supports the use of self-assembled neocartilage toward clinical translation.

For fibrocartilage, studies have applied both dynamic and static tension to the tissue engineering of menisci. For example, dynamic tension at 6% strain was applied to tissue-engineered menisci fabricated from bovine BM-MSCs on poly(ε-caprolactone) nanofibrous scaffolds [492]. Loading improved collagen content and tensile properties with a 16% increase in tensile modulus compared to non-loaded controls. Other studies have applied clamping devices to bovine neocartilage, generated from bovine fibrochondrocytes of the knee menisci embedded in collagen scaffolds; doing so effectively applies passive tension that resists the natural contraction of the engineered tissue [488, 493]. At 6 weeks, tensile moduli reached ~ 1 MPa (for comparison, the tensile modulus of menisci of neonatal calf is 13–26 MPa), while collagen fiber bundles reached ~ 350 μm in diameter, similar to native tissue. While tension has been shown to be useful in hyaline cartilage and fibrocartilage, tension for elastic cartilage tissue engineering is not well investigated.

3.4.9 Hydrostatic Pressure

Hydrostatic pressure applies a uniform normal stress, without any measurable tissue strain (Fig. 3.4) [477]. This loading can be applied using sealed chambers filled with liquid to generate controlled pressure environments. When applied within ranges close to physiological conditions (~3–10 MPa for articular cartilage), hydrostatic pressure has been shown to enhance neocartilage formation by upregulating chondrogenic gene expression and promoting ECM synthesis, particularly of type II collagen and GAGs [477].

The effect of hydrostatic pressure has been investigated in various systems. Hydrostatic pressure magnitude, timing, and combination with growth factors on the mechanical properties of self-assembled neocartilage have been investigated using bovine articular chondrocytes [494, 495]. Static hydrostatic pressure applied from days 10 to 14 of neocartilage maturation at 10 MPa resulted in 96% and 92% increases in aggregate modulus and Young's modulus, respectively. The combination of 10 MPa static hydrostatic pressure and TGF-β1 treatment had an additive effect on aggregate modulus and Young's modulus as well as a synergistic effect on collagen content. In another study, cyclic hydrostatic pressure was applied to bovine articular chondrocytes embedded in a collagen hydrogel for 7 days of application, followed by 7 days of static culture, resulting in increased GAG content [496]. A meta-analysis also suggests that static hydrostatic pressure ranging from 5 to 10 MPa and study lengths of 2 weeks or more were conditions that enabled the detection of significant changes in anabolic response [497]. As shown here, hydrostatic pressure has been applied to enhance cartilage maturation, while applications such as its use for cell differentiation/redifferentiation remain underexplored.

In fibrocartilage, studies have explored the benefits of hydrostatic pressure on fibrochondrocytes from the TMJ disc and knee meniscus. Porcine fibrochondrocytes from the TMJ disc were seeded onto poly-glycolic acid scaffolds, and 10 MPa static or dynamic hydrostatic pressure was applied for 7 days [498]. Dynamic hydrostatic pressure resulted in lower collagen content than the non-loaded control, whereas static hydrostatic pressure resulted in significant increases in collagen content compared to the non-loaded control. In another study, leporine fibrochondrocytes from the knee meniscus were seeded onto poly-L-lactic acid and exposed to either static or dynamic hydrostatic pressure at 10 MPa [499]. In this study, the interaction of hydrostatic pressure with TGF-β1 was also examined. Application of static hydrostatic pressure resulted in fourfold higher collagen content as an additive increase to TGF-β1 treatment alone, threefold higher GAG content as an additive increase, and enhanced compressive properties as a synergistic increase to TGF-β1 treatment. These results suggest that static hydrostatic pressure may be beneficial to both fibrochondrocytes and chondrocytes, while its effect for tissue engineering elastic cartilage remains to be explored.

3.4.10 Shear

Shear stress is a stress component acting parallel to a cross-sectional area of the tissue (Fig. 3.4). In tissue engineering, shear may be applied as fluid-induced shear (e.g., orbital shakers, perfusion bioreactors) or direct mechanical shear (e.g., sliding a solid sphere or platen on the surface). Studies have shown that fluid-induced shear in an optimal range can improve collagen content, tensile properties, and surface frictional properties. Direct mechanical shear typically involves the simultaneous application of a compressive load to mimic the compressive action of articulating joints [500].

The application of shear using bioreactors has been shown to enhance collagen content and tensile properties as well as surface properties. Using primary bovine articular chondrocytes for forming scaffold-free neocartilage and a parallel-plate bioreactor, it was demonstrated that shear stress increased type II collagen content, tensile Young's modulus, and UTS values [501]. A follow-up study demonstrated that increasing shear stress magnitude increased surface layer thickness of type I collagen and tensile properties while decreasing GAG content [502]. Additionally, fluid-induced shear applied to human scaffold-free neocartilage resulted in a 168% increase in tensile modulus without changing GAG and collagen contents [503]. It was suggested that the fluid-induced shear affected collagen crosslinking. In another study, polyurethane scaffolds seeded with bovine articular chondrocytes were subjected to dynamic compression and oscillating sliding motion. The loading improved compressive properties, GAG content, and expression of lubricin on the surface, leading to a 63% reduction in the coefficient of friction compared to the non-loaded control [504]. Thus, shear stress appears to be particularly beneficial in improving tensile and surface properties of neocartilage, possibly through increases in collagen content or crosslinks and surface protein expression.

In fibrocartilage tissue engineering, a few studies have applied shear stress to TMJ disc and knee meniscus-derived fibrochondrocytes. A rotating bioreactor was used to apply shear stress on porcine fibrochondrocytes from the TMJ disc seeded to polyglycolic acid scaffolds; doing so resulted in no clear benefits [505]. In another study, oscillating fluid flow-induced shear stress was applied to fibrochondrocytes from the leporine knee meniscus, resulting in improved GAG content compared to the non-loaded control. Shear stress on elastic cartilage tissue engineering remains to be explored. In summary, application of biomechanical stimuli has the potential to improve neocartilage properties, and further studies combining biochemical and biomechanical factors may further improve the potential for translation.

Chapter concepts:

- Tissue engineering strategies are based on the three pillars of cells, scaffolds or structural frameworks, and biochemical or biomechanical signals.
- Tissue engineering approaches are broadly categorized as scaffold-based or scaffold-free depending on the use of exogenous biomaterials.
- The selection of appropriate cell sources is critical and must consider factors including expansion capacity, ECM production, immunogenicity, ethics, costs, and translational feasibility.
- Autologous, allogeneic, and xenogeneic cells each offer distinct utility and limitations for clinical translation.
- Chondrocytes from various cartilage types have been used in various orthotopic and heterotopic applications.
- MSCs from bone marrow, adipose tissue, synovium, and umbilical cord are widely studied but vary in chondrogenic potential and clinical performance.
- ESCs and iPSCs enable large-scale expansion and fabrication of neocartilage but require tight control of differentiation protocols and face translational hurdles such as tumorigenicity.
- CPCs, isolated from cartilages by methods such as FFA, migration, or cell sorting, exhibit multipotency and strong potential for chondrogenic differentiation.
- Natural and synthetic scaffolds provide structural support and facilitate cellular attachment, proliferation, differentiation, and ECM production for neocartilage formation.
- Composite scaffolds combining natural and synthetic materials may help to recreate the zonal and mechanical complexity of native cartilage.
- Self-organization or self-assembly of cells are scaffold-free methods to form neocartilage without exogenous biomaterials.
- Biochemical cues such as TGF-βs, BMPs, GDFs, and FGFs are essential for directing expansion, differentiation, redifferentiation, and tissue maturation.

- Biomechanical stimuli, including compression, tension, hydrostatic pressure, and shear, enhance ECM production and improve the mechanical properties of neocartilage.
- A combination of biochemical and biomechanical cues may be tailored to the specific cartilage type being engineered to meet its unique structural, mechanical, and functional requirements.

In Vitro Evaluation Methods

4

Rachel C. Nordberg, Takumi Takahashi, Jerry C. Hu,
and Kyriacos A. Athanasiou

To develop design standards for engineered implants and assess how well they recapitulate native tissue, methods are needed to evaluate the functional properties of both native and tissue-engineered constructs. This chapter will describe in vitro assays for assessing the structural and functional properties of native cartilages and tissue-engineered cartilage. Comprehensive in vitro evaluation of an implant, in comparison to the target native tissue it is intended to replace, is an important step before proceeding with in vivo (i.e., animal) studies, which will be described in detail in Chap. 5. Ideally, it is desirable to match the native tissue in terms of structure and function to the greatest extent possible before commencing animal studies. Commonly used evaluation techniques include gross morphology, histology, biochemistry, and biomechanics. In addition, a host of techniques, such as various microscopy and spectroscopy methods and gene sequencing methods, can provide a more nuanced evaluation of cellular and tissue properties. To quantitatively compare the properties of implants to native tissue, a functionality index (FI) can be utilized, which is a weighted assay of multiple assays distilled into a single value. This chapter will describe relevant and emerging technologies for evaluating cartilage tissues and tissue-engineered cartilage.

4.1 Gross Morphology

Gross morphology varies among cartilages. For example, healthy articular cartilage is typically smooth, semi-translucent, and glossy. When developing cartilage implants, it is important to develop constructs that match the morphologic characteristics of native

R. C. Nordberg et al., *Tissue Engineering the Cartilages of the Body*, Synthesis Lectures
on Tissue Engineering, https://doi.org/10.1007/978-3-032-09105-5_4

tissue to facilitate implantation and recapitulate the native tissue structure. Common qualitative assessments that are conducted are visual assessments of shape and the presence or absence of defects or cysts. Quantitative parameters that can be collected include gross morphological measures such as length, width, thickness, and radius of curvature. These parameters may be used in determining how a tissue engineering approach or culture methodology differs from another and also for selecting a methodology for the production of implants intended to repair cartilage defects.

When evaluating tissue-engineered cartilage with the ultimate goal of applying it to cartilage defects, two sets of release criteria may be required, the first being morphologic in nature (e.g., qualitative and quantitative evaluation of gross morphology) and the second being functional in nature (e.g., quantitative evaluation of biochemical and biomechanical parameters) [234]. Failure to satisfy the essential morphological requirements (e.g., flat rather than spherical or folded in geometry) can render neocartilage unsuitable for implantation, irrespective of its other functional attributes. For example, morphology can be heavily impacted by the seeding density used or the stimuli applied during culture. It has been shown that passaging of chondrocytes leads to dedifferentiation and formation of spherical constructs that are not appropriate for implantation [236]. To remedy this, cells may be treated with appropriate biochemical factors and subjected to an aggregate rejuvenation step; these methods restore the chondrogenic phenotype to highly passaged chondrocytes, such that they form flat constructs suitable for repairing cartilage defects [236]. In another example, a necrotic core (i.e., a region with a lack of ECM deposition) has been observed in large cell pellets seeded with a high seeding density [506]; in this case, morphologic assessment can also be used to determine that these pellets are not appropriate for use in addressing cartilage defects. In summary, morphologic assessments can be used as an initial evaluation of tissue-engineered cartilage, and they are often used to determine whether to proceed or not with further development and assessment based on the presence or absence of undesirable morphologic characteristics.

4.2 Histology

Histological evaluation of tissues is commonly used to assess the structural organization of tissues and to provide a qualitative assessment of the composition. For example, overall tissue organization and the distribution of cells, collagen, and proteoglycans can be observed using histology. While histological techniques are primarily qualitative in nature, they can be used in semiquantitative methods as well (e.g., counting the number of cells in a cross-section or quantitatively evaluating the intensity of staining). Histological grading systems have been developed for native tissue and for the evaluation of constructs after in vivo implantation; these will be further discussed in Sect. 5.3.7. In practice, semiquantitative measures used to evaluate histological cross-sections should be further supported through quantitative biochemical assays described in Sect. 4.3.

Commonly used histological stains for evaluating the spectrum of cartilages include H&E, Sirius red in picric acid (picrosirius red), Safranin-O, and Alcian Blue. Overall tissue structure can be evaluated using H&E, where hematoxylin stains the nucleus, which provides information on cellular distribution throughout the tissue, and eosin serves as a counterstain that stains cytoplasmic components and ECM. Often used to study collagen organization, picrosirius red is characterized by a pink or red stain that binds to collagen helices. When imaged under polarized light microscopy, picrosirius red enhances the visualization of collagen fiber alignment, which can provide information on the anisotropic nature of the tissue. However, picrosirius red does not differentiate between collagen subtypes. If more nuanced collagen subtyping information is needed, immunohistochemistry (IHC) can be used to study the distribution of specific proteins such as type I and type II collagens. Safranin-O and Alcian blue are often used to study the proteoglycan organization within cartilage tissue by binding to negatively charged GAGs. While histology can be a useful evaluation technique, it has inherent disadvantages such as being non-quantitative and prone to variations from study to study.

4.3 Biochemistry

The biochemical content of cartilage is frequently measured by spectrometry-based biochemical assays. Common quantitative measures that are reported in cartilage biochemistry testing include collagen content, GAG content, and deoxyribonucleic acid (DNA) content. When reporting results from biochemical assays, there are various ways to represent the data such as normalized by dry weight, wet weight, or DNA content. Inconsistent reporting standards throughout the literature can make comparing results from different publications difficult. Therefore, it is important for researchers to include all relevant data to enable conversions between normalization techniques or to present biochemical data in multiple normalized forms.

Total collagen content is often quantified by a hydroxyproline assay, leveraging hydroxyproline's localization to predominantly collagen [507]. In this assay, hydroxyproline from hydrolyzed cartilage is oxidized by chloramine-T, and Ehrlich's reagent is added to induce a color change based on the degree of oxidation. Collagen has a predictable proportion of hydroxyproline, which allows the collagen content to be estimated by the concentration of hydroxyproline in the sample. Other potential methods for quantifying collagen content include a Sirius red dye binding assay [508] or assays with a fluorogenic reagent binding to the N-terminal glycine residues of collagenase-digested tissue [509]. While these techniques may be useful to determine total collagen, a major disadvantage is that hydroxyproline assays are not able to differentiate between collagen subtypes (e.g., type I vs. type II collagens) that are important for distinguishing among different cartilages, and therefore other techniques are needed to determine the relative quantity of collagen subtypes.

Besides collagen, GAGs are another major structural element of cartilages. GAG content is often quantified by 1,9-dimethyl methylene blue (DMMB) dye binding assays [510]. In this assay, DMMB dye binds to the negatively charged proteoglycans and induces a color change. Unbound DMMB has a peak absorbance of 590 nm, appearing blue, and bound DMMB has a peak absorbance of 530 nm, appearing purple; this change can be quantified by spectrometry methods. Commercially available kits have been developed to improve the stability and ease of use of the originally described assay.

DNA content is also often quantified by biochemical assay techniques such as PicoGreen or Hoechst DNA assays. Both Hoechst and PicoGreen are fluorescent dyes that selectively bind to double-stranded DNA. Mammalian chondrocytes contain approximately 7.7 pg of DNA per cell [511], although this value may vary by cell type and species. Using this conversion, the number of cells can be estimated from the results of the DNA assay. Cellular content within a sample can also be indirectly quantified through metabolic assays such as Alamar blue, which is applied to living cells and changes color as the cells metabolize the non-fluorescent indicator dye. DNA content and metabolic activity can provide insight into the cell growth and activity within a sample.

Mass spectrometry-based proteomics methods have many advantages in the characterization of cartilages. Protein mass spectrometry methods can be used to quantify specific proteins in a targeted manner using techniques such as liquid chromatography-tandem mass spectrometry (LC–MS/MS) or to determine the relative abundance of proteins in a mixture using techniques such as Orbitrap [144]. For targeted mass spectrometry approaches such as LC–MS/MS, the concentration of specific proteins within a sample is determined by separating molecules using liquid chromatography and then performing mass spectrometry on the ionized molecules [11]. This method is not restricted by species-specific limitations that affect antibody-based techniques and allows for the detection of collagen crosslinks. For example, LC–MS/MS was recently used to quantify the ratio of type I and type II collagens and the mature and immature crosslinks within a spectrum of cartilages isolated from the Yucatan minipig [144]. Another mass spectrometry approach is bottom-up proteomic analysis using technologies such as an Orbitrap mass spectrometer. While bottom-up techniques generally cost more to perform and take more time per assay than targeted approaches, they allow for the detection of many proteins from a single trypsin-digested solution using a shotgun approach. The raw files from the Orbitrap run are matched up to a database such as MaxQuant to determine the proteins present within the sample. A limitation is that if data on the species are not available, then the data cannot be analyzed, although databases exist for almost any relevant species currently used in cartilage research (e.g., human, porcine, bovine, murine, etc.). In general, both targeted and bottom-up mass spectrometry approaches are valuable for evaluating native and tissue-engineered cartilage, with the choice between the two methods depending on whether the scientific inquiry focuses on a select set of proteins or requires a broader proteomic analysis.

4.4 Biomechanics

Given that cartilage primarily serves a mechanical role in the body, testing the mechanical properties of tissue-engineered cartilage in comparison to native tissue is of great importance. Mechanical testing of cartilage can broadly be broken down into several categories including compression, tension, tribology, integration, shear, and fatigue.

Compression testing is the most widely used method for assessing the mechanical properties of cartilage [512]. Compression testing can be either displacement-controlled or load-controlled (Fig. 4.1), and samples can be tested in unconfined, confined, or in situ (i.e., locally testing compressive properties of a large tissue sample) configurations. Creep indentation is a commonly used test for obtaining compressive properties of a sample. It applies constant stress to a sample and measures the deformation over time until equilibrium [140, 150]. This test is typically conducted using an open-loop, load-controlled testing apparatus, where a predetermined, constant input load is applied to the sample [140, 150]. In contrast, confined and unconfined compression tests are often conducted on closed-loop, displacement-controlled systems. A typical displacement-controlled test is a stress-relaxation test, where a constant strain is applied to a tissue, and the stress over time is measured until equilibrium. Several other variations of displacement-controlled tests include the ramp tests, which apply a strain at a constant rate while measuring stress, and dynamic tests, which apply cyclic strains to a sample and measure the resulting stress. While bulk mechanical testing is commonly used to characterize cartilage, nanoindentation may be more suitable when spatial resolution is needed (e.g., assessing the compressive properties of different zones of the growth plate) [213]. Nanoindentation uses a nanoscale probe to perform an array of individual mechanical tests and provides localized mechanical property data in a spatially organized manner. Given that tissue properties are dependent on the mechanical testing protocol employed, it is essential that mechanical testing parameters are provided when reporting tissue mechanical properties. Given the diversity of available compression testing methods, when selecting a protocol, it is prudent to consult prior literature and use a widely adopted methodology with established testing parameters to aid in making comparisons across studies.

The tensile properties of cartilage tissue are often studied via uniaxial tension testing. In a strain-to-failure test, samples are cut to a "dog bone" shape, each end is secured to a testing grip, and the sample is subjected to a constant strain rate while the stress is monitored. A stress–strain curve of cartilage will have a toe region where the collagen fibers are straightening out, a linear region where the sample is undergoing linear deformation, a yield point where collagen fibers begin to fail, and a point of structural failure where stress reaches its maximum (Fig. 4.2). Before the yield point, the tissue would return to its original shape if the load were released. At the yield point and beyond, the tissue becomes irreversibly damaged and undergoes plastic deformation. Parameters calculated from the stress–strain curve include Young's modulus (i.e., the slope of the linear region), UTS (i.e., the highest stress observed), strain-at-failure (i.e., the strain corresponding to the

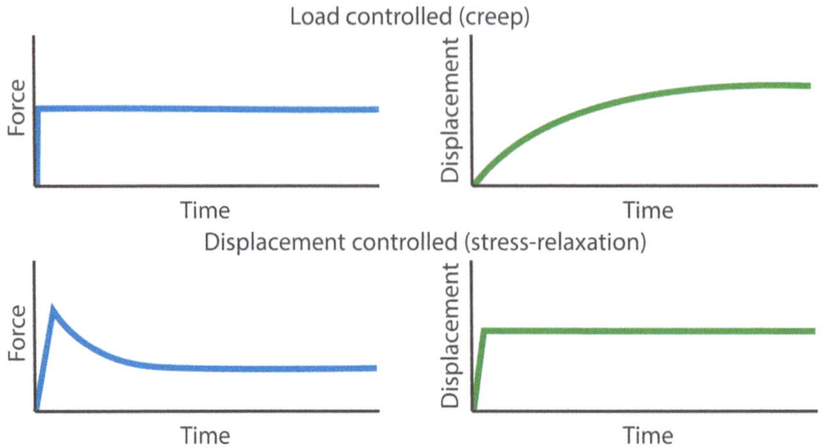

Fig. 4.1 Compression testing regimens. Creep testing is conducted by applying a constant load (top blue line) and measuring the displacement over time (top green line). In contrast, stress-relaxation is performed by applying a defined displacement (bottom green line) and measuring the force dissipation over time (bottom blue line)

highest stress), resilience (i.e., area under the curve until yield), and toughness (i.e., total area under the curve). Besides a strain-to-failure test, tensile testing can also be used in stress-relaxation or cyclic testing configurations by applying a constant or cyclical strain, respectively. Similar to compression testing, the use of standardized testing protocols is needed to allow for inter-study comparisons to be made about tensile properties.

Tribological properties of articular cartilage describe the tissue's ability to glide across another surface and the friction involved in this interaction. Cartilage exhibits several modes of lubrication depending on speed of articulation, magnitude of load, and fluid viscosity, and this relationship can be illustrated in a Stribeck curve, which is shown in Fig. 4.3 along with the corresponding lubrication modes [33]. Briefly, boundary lubrication describes when cartilage surfaces are directly in contact with each other and there is a lack of a fluid film, whereas hydrodynamic lubrication accounts for a fluid film separating the two surfaces. Mixed lubrication has both surface and fluid film interactions, and the elastohydrodynamic lubrication includes fluid film accompanied by cartilage deformation. In vivo, cartilage-to-cartilage coefficient of friction values have been reported as low as 0.001 but often range from 0.1–0.3 in ex vivo boundary lubrication tests [33]. The tribological properties are most commonly determined through pin-on-disc and pin-on-plate configurations [33]. These tests involve sliding a sample submerged in fluid (e.g., PBS) across a surface with a known normal force and measuring the frictional forces, which can be used to calculate the coefficient of friction. Under sufficiently low speeds, these tests are performed under boundary mode lubrication. Surface roughness, measured by

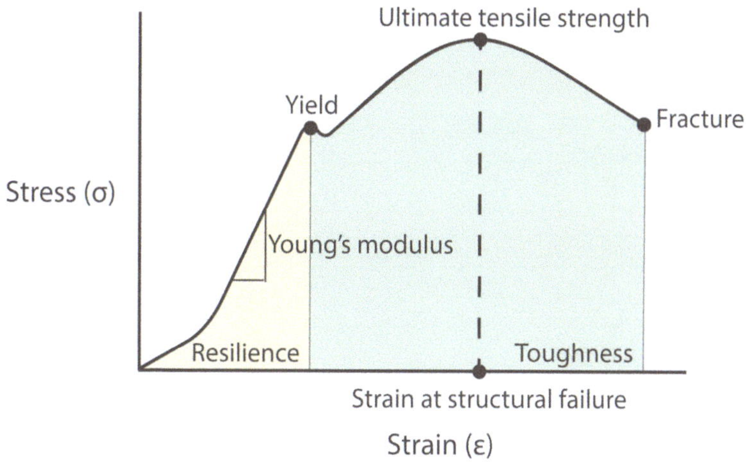

Fig. 4.2 Tensile stress–strain curve. A representative stress–strain curve is shown with key terminology illustrated. The Young's modulus is calculated from the slope of the linear region of the curve. Ultimate tensile strength (UTS) corresponds to the maximum stress observed. Strain at structural failure is defined as the strain at the UTS. Resilience refers to the area under the curve up to the yield point, while toughness is the area under the curve up to fracture

techniques such as interferometry, may also provide insight into the lubrication characteristics of cartilage [180]. Overall, tribological properties are of particular importance for implants intended to replace articulating surfaces.

To assess the integration strength of tissue-engineered cartilage with native cartilage, push-out or uniaxial pull-apart testing methods have been developed. To test integration in vitro, neocartilage is cultured within a defect of a native tissue explant for a period to establish an interface between the two tissues. To assess integration using push-out testing, a probe is then used to push the neocartilage out from the native tissue sample while measuring the resultant stress. Alternatively, the tissues can be trimmed so that one side is tissue-engineered cartilage and one side is native tissue, and this assembly is then tested in uniaxial pull-apart tension tests [513]. While these tests can be performed in vitro, they are also particularly valuable following the in vivo studies described in Chap. 5, as they help assess how well the implant integrates into the defect.

In the case of osteochondral implants, it is also desirable to test the interface mechanics at the cartilage-bone transition. Lap shear testing and uniaxial pull-apart testing have been employed to assess interface mechanics (Fig. 4.4), in which cartilage and bone are affixed to a uniaxial testing setup and sheared apart or pulled apart, respectively. Testing this interface can be technically challenging due to several reasons including the robust mechanics of the osteochondral interface in native tissue, difficulty in securing cartilage to test setups due to its low coefficient of friction, and the risk of cohesive failure (i.e.,

Fig. 4.3 Cartilage lubrication modes. A Stribeck curve illustrates how articulation speed, applied load, and fluid viscosity affect the coefficient of friction (μ). **a** In boundary lubrication, both articular surfaces directly interact without the influence of a fluid film. **b** Mixed lubrication involves characteristics of both boundary contact and the presence of a partial fluid film. **c** In elastohydrodynamic lubrication, a fluid film and load-induced cartilage deformation act in conjunction to influence tribological properties. **d** In hydrodynamic lubrication, the fluid film separates the two articulating surfaces and accounts for all lubrication characteristics. Reproduced with permission [33]

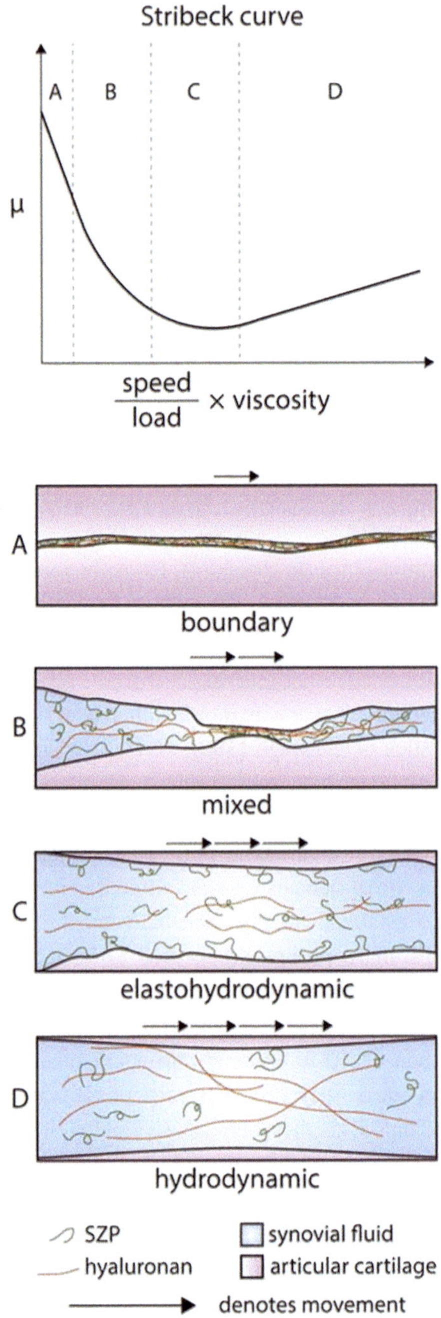

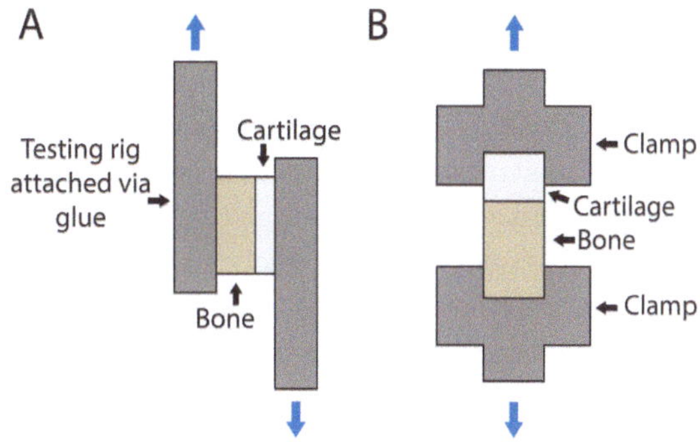

Fig. 4.4 Interface testing configurations. **a** Lap shear testing and **b** pull-apart testing can be used to assess the osteochondral interface mechanics. The axis along which the samples are pulled is denoted by blue arrows. Reproduced with permission [514]

failure within the cartilage) rather than adhesive failure (i.e., failure at the osteochondral interface) [514].

Fatigue testing, while relatively uncommon in cartilage testing, involves repeatedly subjecting a sample to a force (e.g., tensile, compressive, shear) in a cyclic pattern until the sample fails. Fatigue testing gives parameters such as the number of cycles a material can be loaded until failure to provide insight into long-term durability. While this is relatively straightforward in non-living materials, living tissues have additional nuance in that living cells may remodel tissue in response to mechanical stimuli. If the loading regime used is supraphysiologic, the cells may contribute to the degradation of the tissue through the release of catabolic factors. Whereas if the loading is at physiological levels, this may contribute to factors that maintain the ECM.

While there are numerous constitutive models that can be used to study cartilage behavior, some of the ones most often used include the linear elastic model, viscoelastic models, and biphasic theory [38]. The linear elastic solid model is a basic model that assumes that the relationship between stress and strain is linear and is often used to analyze tensile stress–strain data. This is advantageous because it is simple to analyze and it is often accurate enough to describe the tissue under tension if the tissue is tested under slow or quasistatic conditions, but the linear elastic model does not consider any time-dependent material properties. To model time-dependent tissue characteristics, viscoelastic models can be applied to cartilage tissue, such as the standard linear solid model. This model allows for the calculation of parameters such as instantaneous modulus, relaxation modulus, and coefficient of viscosity. More complex models, such as biphasic theory, can be used to consider the composition of the tissue. Specifically, biphasic theory assumes that the tissue has an elastic matrix phase and a fluid phase, and the fluid phase exudes from the matrix upon loading [515]. Biphasic theory is often used to model compression testing and allows for the calculation of the aggregate modulus, permeability, and Poisson's ratio. The reader can find a more in-depth description of the models described here in the book "Introduction to Continuum Biomechanics" by Athanasiou and Natoli, 2008 [516].

4.5 Advanced Techniques for Tissue Visualization

Beyond assessing histological cross-sections with brightfield or polarized imaging, more advanced microscopy and spectroscopy techniques can be used to further assess cellular features, tissue ultrastructure, and tissue composition. While a comprehensive description of all visualization techniques available to researchers is out of the scope of this work, we here provide a high-level overview of some techniques that have been utilized within the tissue engineering field for the characterization of native and tissue-engineered musculoskeletal tissues.

Confocal imaging can be useful when imaging a thick sample, where there is a desire to reduce out-of-plane light or fluorescence [517]. Confocal imaging allows for the optical sectioning of the sample, which enhances the resolution of thick samples. To achieve this, a confocal microscope has a diffraction-limited point toward which the illumination and detection optics are directed, and this point traverses the sample to build an image of the entire region of interest [517]. Overall, confocal imaging is a useful technique for imaging 3D samples such as tissues.

Electron microscopy has been used for the visualization of ultrastructural characteristics of cartilage by scanning a sample with an electron beam and then detecting either reflected or transmitted electrons. In scanning electron microscopy (SEM), the electron beam is applied, and the reflected signal, comprised of secondary, backscattered, and other emitted electrons, is detected [518]. This allows researchers to ascertain information about the surface ultrastructure of a sample and is particularly useful for visualizing information such as fiber alignment or porosity of a sample. For example, SEM has been used to visualize the surface structure of articular cartilage [519], and the porosity of the osteo-portion of tissue-engineered osteochondral samples [520]. Transmission electron microscopy (TEM) detects structures within a sample by using a reactive surface to project an image of the transmitted electrons [518]. TEM can be used to evaluate the internal structure of the tissue, cellular, and subcellular structures (e.g., organelles). Because this technique can visualize subcellular structures, it has been used, for example, to visualize lipid vacuoles in lipochondrocytes [521]. Together, both SEM and TEM can provide complementary, high-resolution information about the ultrastructural properties of native and engineered cartilages.

Raman spectroscopy can be used to probe the structural components of tissue without the use of a label to collect data in a non-invasive or minimally invasive manner [522, 523]. Raman spectroscopy irradiates the sample of interest with a laser, and the interaction with molecular bonds causes a shift in the wavelength of the scattered light. The resulting scatter spectra can be correlated to the biochemical properties of the sample in terms of composition, spatial distribution, and concentration. For example, Raman spectroscopy has been used to study the zonal organization of cartilages and can be used to detect specific components of cartilage structure, such as water, GAG, collagen, cytoplasm, DNA, and apatite by probing specific peaks within the Raman spectra [524]. While Raman spectroscopy is a powerful tool to gain insight into the biochemical composition of a tissue, a limitation is that it can be less sensitive than other methods, such as imaging mass spectrometry (IMS).

IMS can be used to study the biochemical composition of a tissue. For example, matrix-assisted laser desorption/ionization (MALDI)-IMS can be used to probe the presence of biomolecules throughout a sample. In MALDI-IMS, a thin sample is coated in MALDI matrix and is irradiated by a pulsed laser to acquire spectra at discrete locations within the sample [525]. The spectra can be collected in a grid pattern, which allows for spatial mapping of biomolecules within a sample. While MALDI can generate spatial maps of the samples, the resolution is lower than that of Time-of-Flight Secondary Ion Mass Spectrometry (TOF–SIMS). TOF–SIMS is an IMS technique that can be used to detect the surface chemistry of a sample. In this protocol, primary ions are used to bombard the sample of interest, and their impact results in the emission of secondary ions [526]. The secondary ions are collected and sent through a detector system. The speed of the collected secondary ion is related to the mass-to-charge ratio and can, therefore, be resolved within a mass spectrometry detector. Disadvantages of this method are that TOF–SIMS is generally not amenable to quantitative measures, and only ions are detected rather than directly quantifying larger biomolecules within the sample. Within the context of chondral or osteochondral tissue engineering, TOF–SIMS could be a particularly useful method for the detection of calcium ions within the sample. For example, TOF–SIMS has been used to detect the surface chemistry of the matrix and calcium secreted by differentiating human AD-MSCs [527]. Overall, IMS techniques are highly sensitive analytical tools that can be used in conjunction with other assays to provide a comprehensive evaluation of molecular distribution within a sample.

Fluorescence lifetime imaging microscopy (FLIM) can image the spatial distribution of fluorophores within a sample, which can provide insight into metabolic processes, protein–protein interactions, and cellular physiology [528, 529]. In this process, a pulsed laser excites the sample, inducing fluorophore excitation, and the fluorescence decay time of excited fluorophores is measured. This can be done across a sample to provide data in a spatially distributed manner. In applications such as metabolic imaging, autofluorescence within the sample allows this to be conducted in a label-free manner. Therefore, FLIM has the potential to study the metabolic processes within a sample to provide insight into its cellular processes.

Atomic force microscopy (AFM) can be used to generate a topographical map of the mechanical properties of a sample [530]. In this method, a cantilever with a probe is indented into the sample. The deflection of the cantilever and the distance the probe moves allow for the calculation of the sample's mechanical properties. Methods have been developed to perform AFM mechanical tests across the entire topography of the sample by acquiring hundreds of thousands of AFM force curves. AFM enables the generation of a nanoscale map of mechanical properties in structures such as proteins, cell membranes, and tissues.

4.6 Sequencing Technologies

While the techniques described thus far aim to characterize implants on a tissue level, there may also be a need to characterize the native and tissue-engineered samples on a cellular level. Several ribonucleic acid (RNA) sequencing technologies are available that can be employed to gain insight into the gene expression patterns of both native cells and cells used in tissue engineering technologies: bulk RNA-sequencing, single-cell RNA-sequencing (scRNA-seq), spatial RNA-sequencing, and multiomic single-cell chromatin and gene expression assays [531]. Bulk RNA sequencing can be used to detect differentially expressed genes and elucidate mechanisms that are activated within a given cell population. ScRNA-seq can simultaneously sequence thousands of genes from all cells within a cell population, which allows for the detection of heterogeneities and subpopulations of cells. Spatial RNA sequencing (i.e., spatial transcriptomics) maps RNA expression across a tissue section, maintaining spatial relationships. This can be advantageous when engineering tissues with zonal organization, such as the growth plate. In addition to sequencing RNA, new technologies are also available that perform epigenetic characterization of the chromatin structure of the sample at the same time as RNA sequencing, which can provide additional insight into gene regulation mechanisms within a sample. Overall, sequencing technologies can be employed to characterize cell populations and identify key signaling pathways that may be targeted to enhance tissue engineering strategies.

4.7 Other–Omics Methods

In addition to RNA-sequencing technologies (i.e., transcriptomics) and mass spectrometry to interrogate the proteomic composition of a sample (i.e., proteomics), other -omics based methods can potentially be used to interrogate the characteristics of a cartilage sample. For example, lipidomics is an emerging field that uses techniques such as mass spectrometry to probe the lipid composition of a sample. This could have applications in assessing the composition of lipid membranes within a cell population [532]. Additionally, as lipocartilage research is further advanced [521], lipidomic techniques may be an important tool to assess the composition of the lipid vacuoles that are present within lipocartilage tissues. Genomics, which involves sequencing the entire genome of a sample [533], could be considered for the purpose of donor cell selection to ensure there are no major genetic abnormalities in a cell population used for implant production. However, genomic data sets are very large and potentially difficult to interpret because variations in the genome from a given sample do not directly indicate which genes are active. Metabolomics, on the other hand, involves probing the small molecules that are present in a sample, which could provide insight into the cellular physiology of a potential donor cell source [534]. Similar to proteomics and lipidomics, metabolomics is also often carried out using mass spectrometry-based approaches. Overall, high-throughput -omics techniques can provide a comprehensive snapshot of the activity and composition of native and tissue-engineered cartilages and cells.

4.8 Functionality Index (FI)

A tissue-engineered implant should be directly compared to the native tissue it is meant to replace to determine how well it recapitulates native tissue properties. Ideally, comparisons should be made across a range of functional properties, including both biochemical and mechanical characteristics. However, without a method to aggregate the data, it can be challenging to analyze multiple parameters simultaneously. An FI is a weighted average of salient functional properties of tissue-engineered cartilage compared to native tissue [428, 535, 536]. While the parameters included may depend on the objectives of the experiment, they often involve components of both biochemical content (e.g., GAG and collagen contents) and mechanical properties (e.g., compressive or tensile properties). The closer the FI value is to 1, the closer the engineered tissue is to native tissue. The equation of an FI is given below, where GAG, collagen, Young's modulus, and aggregate modulus are included.

$$Functionality\ index = 1/4\left[\left(1-\left|\frac{Col_{Nat}-Col_{Eng}}{Col_{Nat}}\right|\right)+\left(1-\left|\frac{GAG_{Nat}-GAG_{Eng}}{GAG_{Nat}}\right|\right)\right.$$
$$\left.+\left(1-\left|\frac{E^C_{Nat}-E^C_{Eng}}{E^C_{Nat}}\right|\right)+\left(1-\left|\frac{E^T_{Nat}-E^T_{Eng}}{E^T_{Nat}}\right|\right)\right] \tag{4.1}$$

In this equation, Col_{Nat} is the collagen content of native tissue, Col_{Eng} is the collagen content of engineered tissue, GAG_{Nat} is the GAG content of native tissue, GAG_{Eng} is the GAG content of engineered tissue, $E^C{}_{Nat}$ is the aggregate modulus of native tissue, $E^C{}_{Eng}$ is the aggregate modulus of engineered tissue, $E^T{}_{Nat}$ is the tensile Young's modulus of native tissue, and $E^T{}_{Eng}$ is the tensile Young's modulus of engineered tissue. While the FI example given here gives equal weight to the biochemical and biomechanical properties of the tissue, other versions of the FI can incorporate more nuanced biochemical and biomechanical characteristics, such as collagen type, GAG type, coefficient of friction, dynamic properties, etc. The FI is a useful method for quantifying how well tissue-engineered cartilage recapitulates native tissue and is advantageous because it distills multiple functional assays into a single value.

Chapter concepts:

- The gross morphology of tissue-engineered cartilage needs to be appropriate for implantation and can serve as a first selection criterion for evaluating an implant.
- Histologic evaluation can be used to study the tissue architecture of native tissue and tissue-engineered cartilage, but biochemical content should be confirmed with quantitative measures such as GAG and collagen assays.
- There are numerous biomechanical assays that can be used to characterize cartilage tissue, but commonly used techniques include creep indentation for compressive characterization and uniaxial strain to failure testing for tensile characterization.
- Tribological properties of cartilage are dependent on lubrication mode but are often tested in boundary mode lubrication using a pin-on-disc or pin-on-plate configuration.
- There are numerous advanced microscopy and spectroscopy techniques (e.g., confocal imaging, electron microscopy, Raman spectroscopy, IMS, FLIM, and AFM) that can provide additional insight into the structure and function of cartilage tissues.
- Advanced sequencing technologies can provide additional insight into the cellular phenotypes and heterogeneities within native chondrocytes and cells used for tissue engineering purposes.
- An FI can be used to aggregate multiple quantitative functional properties into a single value that can quantitatively describe how well tissue-engineered cartilage recapitulates native tissue.

Animal Models and Preclinical Evaluation Methods

5

Rachel C. Nordberg, Takumi Takahashi, Jerry C. Hu, and Kyriacos A. Athanasiou

Before a tissue-engineered cartilage product can be tested in clinical trials, the safety and efficacy of the implant should be thoroughly tested in preclinical animal studies. Although safety is the primary outcome measure of preclinical studies, they should also establish scientific support for the product (i.e., evidence that the product would be efficacious in humans) [537]. In the United States Food and Drug Administration (FDA) issued "Guidance for industry: Preparation of IDEs and INDs for products intended to repair or replace knee cartilage," the agency recognizes that "there is no perfect animal model of articular cartilage injury" [538]. For load-bearing cartilage repair studies, large animal models such as the swine (e.g., minipigs), goat, sheep, and horse are preferred over small animal models (e.g., rabbits, rodents) [538]. Small animal models do not serve as an appropriate model in terms of cartilage thickness, magnitude of loading, and gait [232, 539]. Additionally, some small animals display a tendency for spontaneous cartilage regeneration in defect models. This section will describe various considerations for performing animal studies including cell sourcing issues and advantages/disadvantages of the various animal models. Additionally, this chapter will build upon Chap. 4's in vitro evaluation methods by describing evaluation methods that are specific to in vivo animal models. This will include non-terminal endpoints (i.e., in-life assays) that can be included with preclinical studies to maximize the data derived from such studies without the use of additional animals and that are analogous to endpoints employed in clinical trials. Although this discussion is relevant to joints such as the TMJ, facet, hip, etc., it should be understood that the majority of the work described below has been conducted in the knee.

R. C. Nordberg et al., *Tissue Engineering the Cartilages of the Body*, Synthesis Lectures on Tissue Engineering, https://doi.org/10.1007/978-3-032-09105-5_5

5.1 Preclinical Cell Sources

Broadly, cell sources can be broken down into three categories: autologous (i.e., isolated from the individual they will be used to treat), allogeneic (i.e., isolated from a different individual of the same species), and xenogeneic (i.e., isolated from a different species). While many of the cell therapy products available on the market today involve the use of autologous cells, the use of an allogeneic cell source can be advantageous in that it avoids challenges associated with an autologous approach, such as donor site morbidity and variability of the implanted cells. Additionally, allogeneic cells can be screened for diseases, sterility, and functionality for a given application [540]. When developing a cell therapy product for clinical use, human cell sources are needed to prevent complications such as immune rejection associated with xenogeneic products. However, a challenge in preclinical studies is that human cells themselves are xenogeneic to animal models. Without the use of immune suppression, human cells will be rejected in animals with competent immune systems. However, immune suppression of animals is not only expensive but may also (1) mask adverse reactions that the preclinical study is examining, (2) prevent the implant from healing into the defect as desired, and (3) be harmful to the health and safety of the animal host. In recognition of these challenges, ACPs, as defined by the FDA [541], may be tested as a surrogate for the human product. ACPs are products that are manufactured using animal cells and are analogous to the human product in terms of cellular characteristics and biologic activity [538].

While an ACP can be either autologous or allogeneic in nature, an allogeneic approach enables the use of a cell bank system, which offers many advantages in the manufacturing of a cell-based product [542–544]. For example, cell banks can be well-characterized and be less susceptible to the biological variability of autologous cell sources. Another advantage of using a cell bank is that it enables off-the-shelf therapeutics and fewer surgeries than the autologous approach. A cell bank system, as described by the FDA, uses a tiered structure (Fig. 5.1) to enable consistent production of product batches over a long period of time [542–544]. The production of a cell bank involves the generation of a "master cell bank" from a given cell source. A single vial from the master cell bank can then be expanded to generate a "working cell bank." Vials from the working cell bank are then used to generate product batches. If a cell bank system is to be used to produce the human clinical product, an analogous manufacturing method of the ACP would also employ a cell bank to produce the preclinical test article [234]. It should be noted that the use of an ACP can bring significant costs to the development of cell-based products. Because of the significant time and money needed to develop, characterize, and manufacture the ACP, the cell source and corresponding animal model should be carefully considered when selecting an animal model. Switching to a different animal model in late-stage development may be challenging due to the need to establish a new ACP and ACP cell bank system.

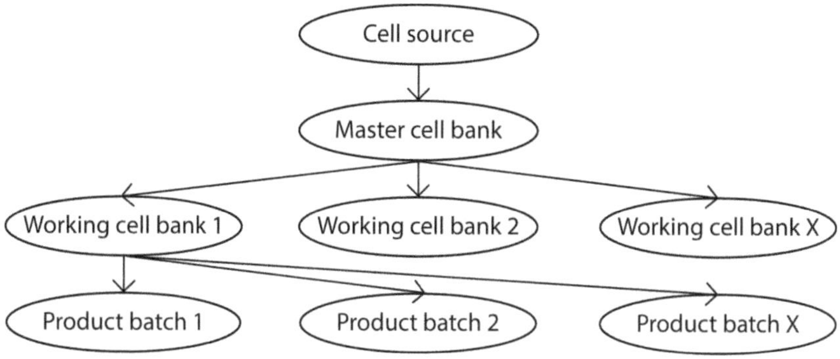

Fig. 5.1 FDA cell bank system structure. The cell banking process starts with isolating cells from a qualified source, which are then processed to establish a master cell bank (MCB). The MCB consists of a uniform collection of vials subjected to comprehensive quality control testing. If the MCB meets all release criteria, a single vial may be expanded to generate a working cell bank (WCB), which is independently tested and qualified. Vials from the WCB are then used to manufacture individual product batches

5.2 Small Animal Models

Following FDA guidance, small animal models are not adequate as models for animal studies intended to show repair of knee cartilage [538]. While specific FDA guidance has not been released for the repair of other diarthrodial joints (e.g., TMJ, hip, facet joint), it is likely that orthopaedic indications would require large animal models, rather than small animal models. Generally, small animal models should be used with caution because cartilage defects have a tendency to spontaneously regenerate, and the anatomy (e.g., cartilage thickness) does not replicate clinical conditions [539]. However, there are cases for the use of small animals, such as conducting mechanistic basic research, conducting studies that require transgenic models, studying non-load bearing cartilage surfaces (e.g., ear), or providing supplementary degradation, durability, and safety data for preclinical studies. This section will describe considerations related to the use of rodent and rabbit small animal models in cartilage repair research.

5.2.1 Rodent

Rodent models (i.e., mouse, rat) are widely used in basic research. Their use in research is driven by substantially lower resource requirements for purchasing, breeding, and housing. Additionally, commercially available strains include both athymic (i.e., immunodeficient) and transgenic models [545]. For example, immunodeficient mice have been used in a number of studies to examine the in vivo maturation of cartilage constructs by implanting the constructs in a heterotopic (i.e., implanted in a location different than its intended clinical location) subcutaneous model. In the early 1990s, Vacanti and associates famously implanted an ear-shaped chondrocyte-seeded polymer scaffold on the back of an athymic mouse model [546]. Numerous other studies have used athymic rodent models to assess in vivo maturation and long-term stability of tissue-engineered cartilage implants [195, 491, 547]. While heterotopic implantation may provide preclinical data for non-loaded cartilages or preliminary toxicology or tumorigenicity data, heterotopic implantation does not recapitulate the loading magnitude and patterns that cartilage experiences in a joint. With regard to orthotopic implantation in a joint (e.g., knee), the mouse has many disadvantages, such as thin cartilage, small joints, and an inability to follow surgical or postoperative procedures that are relevant to the human condition. Therefore, although there are uses for rodent models in basic research and development, it is not an ideal preclinical model for a load-bearing cartilage repair product.

5.2.2 Rabbit

The rabbit model was used in early cartilage repair studies due to its relatively low cost of purchase and care [545]. New Zealand white rabbits were used because defects could be made up to 4 mm in diameter, which was thought to better replicate the human condition [545]. However, a major inadequacy of the rabbit animal model is that cartilage has been documented to spontaneously regenerate from MSCs in the subchondral bone [548]. Because cartilage does not spontaneously regenerate in the human, the process of cartilage repair in the rabbit lacks analogy to the human condition. Moreover, the ability of cartilage to regenerate confounds the results and conclusions of all studies conducted in the rabbit model. That is, a regenerated cartilage surface in the defect site may have been elicited by endogenous healing mechanisms and not necessarily due to the implant or treatment. To further complicate the use of the rabbit model, cartilage thickness in rabbit knees is approximately 0.3–0.4 mm in the knee, substantially lower than in the human [545]. The rabbit knee also has different loading patterns than in the human or large animal models due to a high degree of knee flexion [545]. Due to these concerns, the rabbit has been deemed an insufficient model for the preclinical evaluation of cartilage repair.

5.3 Large Animal Models

Large animal models are the gold standard for cartilage repair because they more accurately replicate the human anatomy, surgical procedures, and mechanical environment than small animal models. It should be noted that while it may seem that a non-human primate model would be most appropriate for a cartilage repair study due to anatomical similarities to the human, non-human primate models are generally not routinely used in cartilage repair due to ethical considerations, high costs, limited availability, and risk of communicable disease transmission to researchers. Rather, consensus standards developed by the American Society for Testing and Materials (ASTM), the International Cartilage Regeneration & Joint Preservation Society (ICRS; formerly known as International Cartilage Repair Society), and the FDA specifically recognize quadruped models such as the pig, goat, sheep, horse, and dog as sufficient to examine preclinical safety and efficacy of cartilage repair therapies [538, 549, 550]. While many cartilage repair studies are conducted in the knee, quadrupedal models are also used for studies in other anatomical locations such as the TMJ [235], hip [551], and spine [552]. This section will describe relevant quadrupedal models and their use within cartilage repair studies.

5.3.1 Pig

Swine, in particular minipigs, are a useful preclinical animal model for cartilage repair due to their comparable size, immune response, genetic makeup, and physiology to the human [553–555]. Within musculoskeletal research, the pig has been recognized as a prominent model for biomechanics research and tissue engineering studies [554]. In particular, minipigs (e.g., Yucatan, Göttingen) are used throughout literature because their size is comparable to that of the human and they are more docile than farm pigs. In addition, minipigs are bred specifically for research purposes, which is advantageous for designing an experiment with a homogeneous population of animals to reduce the effect of biologic variability [555]. The minipig has emerged as a premier model for TMJ-related research, including both disc [235] and condylar studies [556]. Porcine models have also been used in meniscal repair studies [557], and are recognized by the FDA and ASTM as an appropriate model for spine-related indications [552, 558]. The pig has anatomy that replicates the human condition. For example, in terms of knee anatomy, porcine models have a cartilage thickness in the knee that ranges from 1.5 to 3.2 mm, which is comparable to that of humans [555]. The pig has a passive range of motion that ranges from 42° to 144°, which is comparable to other quadrupedal large animal models [555]. The porcine models can be housed in large animal facilities, and there are good laboratory practice (GLP) facilities that conduct porcine research [555]. An advantage of using the minipig is that temperamentally, they are relatively docile upon awaking from anesthesia, thereby enabling human-like postoperative procedures. For example, the minipig is amenable to the use of a postoperative sling to prevent loading of the joint immediately upon awakening from surgery [559]. Overall, the porcine model is recognized as a valuable model for the evaluation of the safety and activity of cartilage repair products.

5.3.2 Goat

Goats are another large animal model that has been mentioned by the FDA as a potential preclinical model for cartilage repair studies. Similar to the minipig, goats have a passive range of motion in the knee, ranging from 45° to 146° [555]. However, cartilage thickness in the knee is thinner than the minipig and the human, ranging from 0.7 to 1.5 mm [555]. Goats require similar large animal facilities to the minipig and may be less expensive to purchase [545]. However, there are two distinct drawbacks to using the goat as a large animal model. First, the goat is prone to subchondral cyst formation and structural collapse of the subchondral bone following chondral or osteochondral defect creation in the knee [560]. Since subchondral cyst formation is not witnessed in human patients undergoing cartilage repair treatments, this indicates that the healing response in the goat is not representative of the human situation. Moreover, goats have a tendency for a flight response upon awaking from anesthesia and may jump around, thereby damaging newly implanted neocartilage [555]. Human patients undergo a recovery phase before returning to full physical activity, and, therefore, the postoperative protocols for goats do not replicate the human condition. Overall, while the FDA has stated that goats are an appropriate model for cartilage repair studies, these limitations should be considered when designing goat studies.

5.3.3 Sheep

Another ruminant that has been named as a potential large animal model for cartilage repair by the FDA is the sheep. Many characteristics of the sheep model are similar to those of the goat. For example, sheep have a similar knee passive range of motion to goats and minipigs, ranging from 40° to 147° [555]. Similar to the goat, the cartilage thickness in the knee is thinner than in the human, ranging from 0.4 to 1.2 mm [555]. The limitations of the goat model are also persistent within the sheep model. For example, the sheep model is also prone to subchondral cavitation in the knee [561], indicating that different healing mechanisms may be present within this model than in the human. Additionally, they are also prone to a flight response upon awaking from surgical protocols [555]. Therefore, many of the same disadvantages of the goat are also present in the sheep model.

5.3.4 Horse

The horse is the largest animal model that is used in cartilage repair studies, weighing up to 500 kg [545]. The horse has been cited by the FDA as an appropriate model for cartilage repair studies [538], and it has been used successfully for translating a cartilage repair product to the clinic. Specifically, an equine model was used for MACI's pivotal large animal study [562]. Musculoskeletal research and treatments within the equine model are fairly well developed due to the horse racing industry [545]. Horses also have naturally occurring conditions related to cartilage degeneration, such as osteochondritis dissecans, cartilage injury, and OA [545]. As companion animals, they may be enrolled in veterinary clinical trials, which can provide proof-of-concept data to support the safety and efficacy of a potential cartilage repair product [563]. In terms of anatomy, the horse replicates the human condition fairly well, with a cartilage thickness ranging from 1.5 to 3 mm in the knee, which is comparable to that of the human. Additionally, it has been reported that the passive range of motion of the knee is 30° to 160°, which is similar to other quadrupedal models [555]. However, unless researchers are located at specific institutions with equine expertise and infrastructure, horses can be challenging to use in preclinical studies because they require equine-specific facilities, which are less common than large animal facilities that are used to house pigs and ruminants. Additionally, awaking from anesthesia can be a challenging and precarious event for the horse due to its large size and tendency to attempt to flee [555]. Therefore, extra precaution, resources, and expertise are needed to successfully implement an equine model for preclinical research.

5.3.5 Dog

The canine model has several advantages over the goat, sheep, and horse. Dogs are smaller in size enabling easier handling, can be trained, and are amenable to postoperative procedures that are relevant to the human situation [555]. However, as companion animals, there are ethical concerns with the use of canine models within biomedical research, and there has been an effort to reduce their usage [545]. Similar to horses, dogs may be enrolled in veterinary clinical trials, which can help provide preliminary safety and efficacy data to support the pursuit of a clinical product [563]. In terms of anatomy, the dog has a similar passive range of motion to other quadrupedal models (34° to 160°) and knee cartilage thickness that is on the lower end of quadrupedal large animal models (0.95 to 1.3 mm) [555]. Overall, the canine model may be a useful model to consider, particularly within the context of veterinary clinical trials.

5.4 Preclinical Evaluations

When designing animal studies, in addition to evaluating the native tissue, implant, and repair tissue using methods described in Chap. 4, non-terminal endpoints and clinically inspired evaluation methods should be incorporated into the experimental design. Non-terminal endpoints have several important advantages such as longitudinal tracking of the healing and immune responses, increasing the data that can be collected from fewer animals, and having analogy to clinical outcome measures that would be collected in human clinical trials. Non-terminal evaluation methods can serve as early indications of whether or not a treatment is safe and/or effective. Notable non-terminal endpoints that have been utilized in animal studies include arthroscopic methods in the knee, blood panels, pain scales, imaging techniques, gait analysis, and activity monitoring. Once an in vivo study has concluded, in addition to the evaluation of tissue using methods described in Chap. 4, histological sections can be evaluated using clinical scoring systems for cartilage degeneration and repair. This section will describe evaluation methods that are relevant to preclinical studies including non-terminal methods and terminal methods that are specifically relevant to in vivo studies.

5.4.1 Arthroscopy

Arthroscopy involves making a small incision into the joint (particularly the knee) through which procedures can be conducted in a minimally invasive manner. There are several types of arthroscopic procedures that may be carried out including diagnostic, tissue removal or repair, deployment of a therapeutic, or second look protocols to assess the efficacy of a treatment. Although arthroscopic procedures require the joint capsule to be opened through a small incision, they preserve the structure of the joint better than open joint procedures. There is great clinical interest in developing novel therapeutics that can be deployed via arthroscopy. Beyond using arthroscopic methods to conduct surgical interventions, second-look arthroscopy can be used to longitudinally evaluate the degree of tissue repair in a minimally invasive, non-terminal manner. This is performed by inserting a camera into the joint to conduct a macroscopic assessment of the tissue. For example, second-look arthroscopy has been described for the longitudinal evaluation of cartilage repair in the minipig knee [559]. Additionally, it has been suggested that arthroscopy is possible in models such as the sheep [564] and horse [565]. Although considered minimally invasive, arthroscopy generally requires that animals be anesthetized. Thus, while incorporating second-look arthroscopic evaluation of cartilage repair may provide researchers with an additional method for evaluating tissue repair without the need to terminate a study, these procedures nonetheless add costs and risks to the animal study.

5.4.2 Blood Panels

Blood panels can be a useful tool to evaluate the safety of an implant. For example, a complete blood count (CBC) is commonly used; this assay quantifies parameters regarding white blood cells, neutrophils, lymphocytes, monocytes, eosinophils, basophils, red blood cells, hemoglobin, hematocrit, platelets, and the presence of clots. Additionally, blood chemistry panels of the plasma are also often taken to quantify the concentration of blood components such as alanine transaminase, aspartate transaminase, albumin, alkaline phosphatase, amylase, blood urea nitrogen, calcium, chloride, creatinine, glucose, potassium, sodium, phosphorus, total bilirubin, total protein, and hemolysis. To collect samples for these panels, a blood draw is performed on an animal, and blood is then placed into either vials with ethylenediaminetetraacetic acid (EDTA) for a CBC panel or vials with no anticoagulant for a blood chemistry panel that are subsequently spun down to separate the serum for analysis. The results from these assays are often compared against reference ranges that are specific to the animal model of interest (e.g., species, sex, age). An advantage of blood panels is that blood can be drawn in a non-terminal manner. Thus, a series of blood draws can be taken from the same animal (e.g., baseline before implantation vs. after implantation). Another advantage is that blood panels are analogously conducted in the clinical population. The disadvantages of blood panels are that they can be difficult to obtain a sample from a conscious animal, it is an invasive protocol that involves puncturing the skin, there can be high standard deviations in some of the parameters quantified, and there may not be reliable reference values available for the specific animal model used. Nevertheless, blood panels are generally considered a useful tool for assessing the safety of an implant.

5.4.3 In Vivo Imaging

In human clinical trials, imaging modalities are often used to assess disease progression and the success of an intervention, and, therefore, preclinical studies may benefit from using analogous imaging techniques to assess cartilage tissue repair. Common imaging techniques used to assess implanted tissues include radiography, magnetic resonance imaging (MRI), and computed tomography (CT). For many of these imaging techniques, scoring systems have been developed, for example, to assess OA progression. These techniques and scoring systems have often been adapted for use in studies that assess cartilage repair in a defect model.

Conventional radiography (i.e., x-ray) allows for the visualization of bone structure and provides an indirect measure of cartilage health [566]. For example, the Kellgren and Lawrence scale was developed to assess joints for OA by radiographic means such as evaluating for evidence of osteophytes, periarticular ossicles, joint space narrowing, altered shape of the bone, and subchondral sclerosis [567]. Based on the evaluation of these qualities, the joint can be classified within five grades of degeneration: (0) none, (1) doubtful, (2) minimal, (3) moderate, or (4) severe. A major disadvantage of this technique and scoring system is that the cartilage cannot be directly visualized, but, rather, it relies upon structural changes within the bone to detect cartilage pathology.

Using MRI to assess the development of OA and the efficacy of cartilage repair in the knee has advantages in that it allows for the direct visualization of the cartilage tissue and other tissues that are involved in joint mechanics (e.g., joint capsule, ligaments). Various MRI-based scoring systems have been developed to evaluate cartilage degeneration and OA development in the knee including the modified Outerbridge grading system, Whole-Organ Magnetic Resonance Imaging Score (WORMS), Boston-Leeds Osteoarthritis Knee Score (BLOKS), Knee Osteoarthritis Scoring System (KOSS), and MRI Osteoarthritis Knee Score (MOAKS) [566]. MRI can also be used to quantify cartilage thickness or volume as an indicator of OA progression. Techniques such as diffusion MRI and sodium MRI have been used to assess ECM content and alignment [568]. Additionally, the Magnetic Resonance Observation of Cartilage Repair Tissue (MOCART) scoring system assesses the degree of defect repair and filling of the defect, integration to the border zone, surface of the repair tissue, structure of the repair tissue, signal intensity of the repair tissue, subchondral lamina, subchondral bone, adhesions, and synovitis [569]. The MOCART system has been used to assess the success of cartilage repair procedures such as MACI [570], and is particularly relevant to assessing the success of future tissue engineering strategies for cartilage repair in both preclinical and clinical settings. However, a major disadvantage of MRI is that it is relatively expensive, and, therefore, may not be feasible to include within preclinical studies.

Similar to radiography, CT-based techniques do not provide direct visualization of the cartilage structure, and, therefore, primarily provide information on the bony structure of the joint. However, contrast agents can be applied via intra-articular injection to enhance the ability to visualize cartilage using CT [566]. In the context of preclinical animal studies, CT scans have been used to assess repair tissue formation, especially in the context of osteochondral approaches, and to assess subchondral bone remodelling [571, 572]. CT scans are relatively inexpensive, and the equipment is often more accessible than MRI. As a result, CT remains a widely used imaging modality in early-stage preclinical investigations.

5.4.4 Pain Scales

Clinically, pain reduction is often the key determinant of patient satisfaction with a given treatment, but quantifying pain in an animal model can be challenging. Small animal models have several types of pain quantification methodologies that have been developed including grimace scales, quantification of behavior characteristics, and external stimulus-based tests [573, 574]. For example, the Von Frey test uses mechanical perturbation to determine the threshold that elicits a paw withdrawal response in rodents [573]. In large animals, there have been pain scales developed for evaluating pain in farm animals, such as the Unesp-Botucatu Sheep Acute Composite Pain Scale, Unesp-Botucatu Pig Composite Acute Pain Scale, Sheep Pain Facial Expression Scale, and Piglet Grimace Scale [575].

Canine models have also used scoring systems such as the Canine Brief Pain Inventory and Canine Research Orthopedic Pain Inventory, which can have the added benefit of being completed by the dog's owner who is very familiar with the behavioral characteristics of the individual dog. For example, the Canine Brief Pain Inventory was used to detect OA-related pain in dogs that were treated with carprofen or a placebo by having the owner complete the survey at the beginning and after 14 days of treatment [576]. Additionally, it has been demonstrated that the Canine Research Orthopedic Pain Inventory scale can be used to assess cartilage repair in a canine model [551]. While animal pain scales have yet to reach widespread use in preclinical cartilage repair studies, researchers should consider incorporating them into future studies to gather non-terminal data and increase the analogy of the preclinical study to a clinical trial, which will likely include patient-reported pain data.

5.4.5 Gait Analysis

Gait analysis can be used to determine whether a treatment impairs or restores the ability of an animal to ambulate satisfactorily. Gait analysis of rodent models has been conducted extensively in literature and has been largely standardized using scoring systems such as the Basso, Beattie, and Bresnahan (BBB) Scale, and specific forced walking tests such as the CatWalk treadmill or ladder rung test [577]. However, in the context of cartilage repair, where large animal models are needed, there are fewer studies that have developed methods to assess gait parameters. Nevertheless, several qualitative and quantitative methods have been developed to assess gait in large animals such as the pig or sheep in large part due to efforts within the field of neurological injuries [577].

Qualitative scoring systems have been developed to assess the gait patterns in large animal models. For example, the Thoracic Injury Behavior Scale (TIBS) scoring system has been used to evaluate gait patterns in the sheep, and a separate Porcine Thoracic Injury Behavior Scale (PTIBS) scoring system has been developed for use in the porcine model [577]. In these scoring systems, researchers review photos or videos of the animals and rate them on a scale of 1 to 10, ranging from no hindlimb movement to normal ambulation. Beyond the use of subjective scoring systems, several quantitative methods have also been developed, such as using pressure-sensing mats or image-based methods of tracking the kinematic movement of animals and analyzing the movement through specialized software [577]. For example, kinematic analysis of the canine model walking on force plates allowed researchers to calculate peak vertical force and vertical impulse as surrogate measures of lameness [551]. Additionally, they quantified the distribution of body weight as the animals recovered from a cartilage resurfacing procedure in the hip joint to show whether the animals avoided using the operated limb [551]. Gait tests can be carried out with different setups, such as monitoring animals as they freely walk around their pen, guiding them through straight-line walking, or using treadmills to modulate walking speed [577]. Overall, gait analysis methods for large animal studies have proven to be feasible, but more standardization is needed, and there may be a need to develop screening systems that are specific to cartilage repair products rather than directly using the tests described above for spinal cord injury.

5.4.6 Activity Monitoring

The activity of an animal can serve as a surrogate measure for pain and physical ability. While the use of activity monitors is not yet widespread in large animal studies, several recent studies have reported using activity monitoring protocols in large animal models. For example, the recovery of a canine model from a hip resurfacing procedure was tracked by affixing an accelerometer to the collar of each dog to track the translational movement of the animal [551]. Another method of activity tracking has been to record the number of steps an animal takes using a Fitbit or similar device. For example, one group attached a Fitbit to the harness of a Yucatan minipig model and extracted the number of steps taken by each animal using the Fitbit application [578]. Using an activity monitor to track motion or step number can be a useful method to quantitatively compare activity to presurgical levels or among experimental groups.

5.4.7 Adaptation of Clinical Histological Evaluation Scales to Animal Models

Once the preclinical study has been terminated and the tissue has been isolated from the animal model, histological scoring systems originally developed from human clinical data can be used in the preclinical setting. The use of standardized scoring systems can provide semiquantitative insight into the degree of cartilage degeneration or repair. For example, the Osteoarthritis Research Society International (OARSI) cartilage OA histopathology grading system was developed to assess the degree of cartilage degeneration within a specimen in relation to how far OA has progressed [579]. The Modified O'Driscoll Scale (MODS) was adapted to rabbit models to assess cartilage tissue repair [580]. However, the requirement by MODS for whole-joint visualization limits its clinical applicability. In an effort to develop a system that can be used to assess clinical biopsies, the ICRS Visual Assessment Scale II (ICRS II) was developed [581]. ICRS II assesses cartilage repair for 14 parameters: tissue morphology, matrix staining, cell morphology, chondrocyte clustering, surface architecture, basal integration, formation of a tidemark, subchondral bone abnormalities, inflammation, abnormal calcification, vascularization, superficial assessment, mid/deep zone assessment, and overall assessment [581]. It should be noted that taking a biopsy of repair tissue can predispose the tissue to further degeneration, and, therefore, using these scoring systems in preclinical studies may not be directly analogous to clinical studies, which typically do not yield histological specimens for evaluation.

Chapter concepts:

- ACPs are animal cell-based analogs to the human clinical product that can be used in preclinical studies to test the safety and efficacy of a product.
- Tiered cell bank systems can be used to manufacture consistent product batches over a long period of time from a single cell source.
- Small animal models (e.g., rodent, rabbit) are generally not sufficient to test the safety and efficacy of load-bearing cartilages in preclinical studies.
- The FDA specifically calls for large animals (e.g., pig, goat, sheep, horse, dog) to test the safety and efficacy of cartilage products intended to repair the knee. This guidance is likely to be applicable to other load-bearing cartilage indications.
- The minipig has emerged as a valuable model for cartilage repair in multiple anatomical locations, such as the knee and TMJ.
- Non-terminal endpoints such as arthroscopy, blood panels, imaging techniques, pain scales, gait analysis, and activity monitoring may reduce the number of animals needed for intermediate time points and provide safety and efficacy data that are analogous to those to be collected in human clinical trials.
- Many preclinical evaluation techniques, such as arthroscopy and ICRS scoring, are specifically developed for the knee; additional work may be required to develop these protocols to be applicable to cartilages in other anatomical locations.

Translational Considerations 6

Takumi Takahashi, Rachel C. Nordberg, Jerry C. Hu,
and Kyriacos A. Athanasiou

Cartilages throughout the human body present distinct scientific and clinical challenges for academic researchers. Furthermore, regardless of cartilage type, successful translation from benchtop to bedside requires a comprehensive understanding of considerations that extend beyond those of academic research [582]. One of the foremost is understanding and navigating the regulatory pathway under the jurisdiction where the product is intended to be marketed. For cell-based, tissue-engineered cartilage products, this pathway typically involves conducting preclinical studies that meet regulatory standards, receiving authorization for first-in-human trials, completing a series of clinical studies to demonstrate safety and efficacy, obtaining marketing approval, and fulfilling post-marketing surveillance requirements to maintain licensure. At each of these stages, researchers must adhere to various regulations and guidelines for good practices, collectively referred to as GxP (Good [x] Practice, with "x" being Laboratory, Manufacturing, Clinical, etc.). Intellectual property (IP) considerations are also essential in navigating the translational pathway because patents help de-risk the technology and improve prospects for commercialization. IP is especially important for attracting financial investments needed to support research and development, preclinical studies, clinical trials, and manufacturing. Another challenge lies in securing the necessary resources, including funding and access to expertise. These resources may include financial support for translational efforts and collaboration with key stakeholders such as university technology transfer offices (TTOs), patent lawyers, venture capitalists (VCs), business professionals, regulatory consultants, clinicians, and hospital personnel. A holistic understanding of the translational process is therefore necessary to successfully bring a tissue-engineered cartilage product from benchtop to bedside. This chapter covers these translational considerations from the perspective of an academic researcher, using the United States regulatory framework as an example.

R. C. Nordberg et al., *Tissue Engineering the Cartilages of the Body*, Synthesis Lectures on Tissue Engineering, https://doi.org/10.1007/978-3-032-09105-5_6

6.1 Regulatory Landscape

6.1.1 The United States Food and Drug Administration (FDA)

The FDA is the primary regulatory authority overseeing tissue-engineered cartilage products in the United States. The FDA is organized into multiple centers, each responsible for regulating different categories of medical products (Fig. 6.1). The Center for Biologics Evaluation and Research (CBER) is responsible for regulating cell-based therapies, including most tissue engineering products. The Center for Devices and Radiological Health (CDRH) oversees medical devices, which may include scaffolds or biomaterials. The Center for Drug Evaluation and Research (CDER) evaluates pharmaceutical drug products, which may intersect with tissue engineering when bioactive molecules or growth factors are included in the product. Any combination of a biologic, device, or drug may be referred to the Office of Combination Products, which assigns primary jurisdiction to one of the three centers based on the primary mode of action of the product. Typically, CBER is assigned the primary jurisdiction for cell-based, tissue-engineered cartilage products. Understanding the distinctions between these centers is important for product developers because the classification of a product determines its regulatory pathway and the type of preclinical and clinical data required for approval.

6.1.2 The Evolution of Regulatory Frameworks

The regulatory landscape for cell-based products has evolved significantly over the past several decades and continues to evolve as emerging technologies challenge the existing framework [231]. In 1993, the FDA published the "Points to Consider in the Characterization of Cell Lines Used to Produce Biologicals," clarifying that cell-based therapies would be regulated as biologics that typically fall under the jurisdiction of CBER [582]. This document outlined that biologics would require preclinical studies, Investigational New Drug (IND) application submission for approval to conduct first-in-human studies and multi-phased clinical trials, and Biologics License Application (BLA) submission for market approval. In 1997, the FDA approved Genzyme's Carticel (autologous cultured chondrocytes), the first cell-based therapy for cartilage repair approved in the United States, a landmark for cartilage tissue engineering [583]. Carticel's approval in 1997, under a full BLA, demonstrated that culture-expanded cartilage-derived cells would be treated as more-than-minimally manipulated cells. This was an early case of a tissue-engineered cartilage product navigating the regulatory pathway in the United States.

By the early 2000s, the FDA introduced a comprehensive, risk-based framework for human cell, tissue, and cellular-and-tissue-based products, commonly referred to as HCT/Ps. In 1997, the agency proposed a tiered approach to regulate all HCT/Ps, aiming to "exercise regulatory oversight only to the degree appropriate to protect public health"

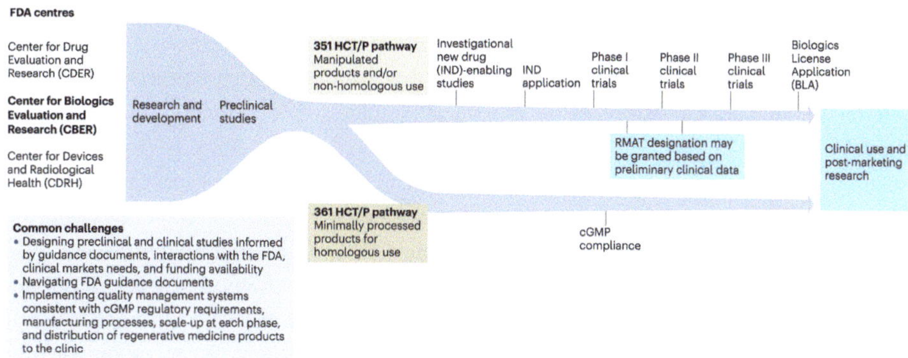

Fig. 6.1 Structure of the FDA approval process. In the United States, cell-based products, including tissue-engineered cartilage products, are primarily overseen by the Center for Biologics Evaluation and Research (CBER) within the United States Food and Drug Administration (FDA). Products that are minimally processed and intended for homologous use may be regulated under the 361 pathway for human cells, tissues, and cellular and tissue-based products (HCT/Ps). However, most cell-based products are considered more than minimally manipulated and regulated under the more stringent 351 HCT/P pathway. After completing preclinical and IND-enabling studies, a sponsor may submit an Investigational New Drug (IND) application to CBER to initiate clinical trials. If early clinical data are promising, the therapy may qualify for a Regenerative Medicine Advanced Therapy (RMAT) designation, which can facilitate expedited development. Upon successful completion of clinical trials, a Biologics License Application (BLA) is submitted to obtain FDA approval for market authorization. Key regulatory challenges include designing appropriate preclinical and clinical studies, interpreting and applying FDA guidance documents, and establishing robust quality management systems. Compliance with current Good Manufacturing Practice (cGMP) regulations is also essential throughout development. Reproduced with permission [582]

[584]. The regulatory framework for HCT/Ps was formalized with the establishment of 21 Code of Federal Regulations (CFR) Part 1271, implemented in phases between 2001 and 2005 [585]. This rule created a unified, risk-based system for regulating HCT/Ps, including tissue-engineered cartilage products. It introduced requirements for registration and listing (2001), donor eligibility (2005), and current Good Tissue Practices (cGTP, 2005). The framework also established a two-tiered pathway, distinguishing between Section 361 HCT/Ps, which do not require premarket approval, and Section 351 HCT/Ps, which are regulated as biologics and require IND and BLA approvals. This regulatory structure continues to guide the development and oversight of tissue-engineered cartilage therapies in the United States.

Various guidance documents have been issued by the FDA over the years that may pertain to tissue-engineered cartilage products. "Guidance for Industry: Preparation of IDEs and INDs for Products Intended to Repair or Replace Knee Cartilage," finalized in 2011, is currently the only document specific to cartilage [233]. While specific to knee cartilage, this document is applicable to products that intend to replace cartilages other than knee cartilage. Other guidance documents, finalized or in draft, include those that cover preclinical assessment, design of early-phase clinical trials, manufacturing changes and comparability, potency assurance, use of human-and animal-derived materials in manufacturing, and safety testing of human allogeneic expanded cells [537, 586–590]. Numerous other guidance documents are applicable if gene therapy is involved in the tissue-engineered cartilage product [591–593]. It is important to consult these guidance documents as applicable to the specific tissue-engineered cartilage product, stay informed of updates, and understand how the FDA interprets and applies them.

International standards recognized by the FDA also play an important role in guiding the development of tissue-engineered cartilage products, particularly in areas not fully addressed by FDA-issued guidance documents. The International Organization for Standardization (ISO) develops globally accepted standards that ensure quality, safety, and consistency across industries. Relevant ISO standards include ISO 13485 (quality management systems), ISO 11737 (sterility assurance), ISO 14644 (cleanroom classification), ISO 14971 (risk management), ISO 20399 (biobanking), ISO 10993 series (biological evaluation), and ISO 21973 (transport of therapeutic cells) [594–600]. The International Council for Harmonisation (ICH) issues guidelines specifically for pharmaceuticals and biologics, including ICH Q5A(R2) (viral safety), Q5C (stability testing), Q6B (specifications), S6(R1) (preclinical safety), and E6, E8, and E9 (clinical trial design and conduct) [601]. ASTM International provides consensus-based standards focused on testing methods and material properties, such as ASTM F2027 for characterization and testing of raw or starting materials for tissue engineering and ASTM F2150 for characterization and testing of biomaterial scaffolds for tissue engineering [602, 603]. While not a comprehensive list, these examples highlight the importance of consulting both international standards and local regulatory guidance when developing tissue-engineered cartilage products.

Regulatory updates may also accelerate the approval process of tissue-engineered cartilage products. The 21st Century Cures Act, signed into law in December 2016, marked a major legislative milestone for the advancement of regenerative medicine in the United States [604]. Among its key provisions was the creation of the Regenerative Medicine Advanced Therapy (RMAT) designation, a program specifically designed to expedite the development and review of regenerative therapies [582]. The RMAT designation applies to products that meet the FDA's definition of a regenerative medicine therapy ("a cell therapy, therapeutic tissue engineering product, human cell and tissue product, or any combination product using such therapies or products, except for those regulated solely under Section 361") that are intended to treat, modify, reverse, or cure a serious or life-threatening disease or condition [605]. To qualify, the product must demonstrate preliminary clinical evidence indicating its potential to address unmet medical needs. Therapies granted RMAT designation receive benefits similar to those of Breakthrough Therapy designation, including early and frequent interactions with the FDA, priority review, and eligibility for accelerated approval based on surrogate or intermediate clinical endpoints. To date, RMAT designations have been granted for indications such as knee OA and chronic low back pain associated with degenerative disc disease [582]. For developers of tissue-engineered cartilage products, obtaining an RMAT designation offers a potential pathway to engage with regulators early in development, streamline clinical trial design, and shorten the time to market.

6.1.3 Regulatory Pathways for Cartilage Products

In the United States, HCT/Ps are regulated under the aforementioned two distinct pathways defined by the Public Health Service (PHS) Act (Fig. 6.1) [606]. Products that are minimally manipulated and intended for homologous use (meaning "the repair, reconstruction, replacement, or supplementation of a recipient's cells or tissues with an HCT/P that performs the same basic function or functions in the recipient as in the donor") are regulated under Section 361 (361 HCT/P pathway) [607]. This pathway imposes relatively limited regulatory requirements and does not mandate clinical trials for market entry. Several cartilage repair products, such as DeNovo NT, ProChondrix, and Cartiform, have followed the 361 HCT/P pathway because they consist of minimally processed allograft tissue [232]. While the 361 HCT/P pathway may be less financially and technically burdensome, it does not inherently facilitate insurance reimbursement. Consequently, even products that enter the market under the 361 HCT/P pathway may need to generate robust clinical data to improve the likelihood of payer acceptance.

In contrast, products that are more-than-minimally manipulated or used in a non-homologous manner fall under Section 351 of the PHS Act (351 HCT/P pathway). For products regulated under the 351 HCT/P pathway, the initial regulatory milestone is obtaining FDA authorization to begin clinical trials by submitting an IND application, as described in 21 CFR Part 312 [608]. This application must provide sufficient data to support the product's safety and justify its use in human subjects. Sponsors may seek early input from the FDA through optional meetings such as the INitial Targeted Engagement for Regulatory Advice on CBER/CDER ProducTs (INTERACT) or a formal pre-IND meeting. A complete IND submission includes an overview of the investigational plan, investigator's brochure, clinical trial protocols, chemistry, manufacturing, and control (CMC), toxicology, and pharmacology findings from preclinical studies, and any previous clinical experience. Depending on the intended use and target population, additional information may be required, such as plans for pediatric evaluation or responses to specific requests by the FDA.

Establishing a Quality Management System (QMS) is fundamental during IND preparation. This includes adhering to GLP for preclinical studies (pivotal animal studies), which involves well-defined protocols, standard operating procedures (SOPs), reliable data collection, and trained personnel. If certain GLP conditions are not met, sponsors must provide a justification within the IND submission. Once submitted, the FDA reviews the IND within 30 calendar days to determine whether the proposed clinical study may proceed.

Upon IND authorization, human clinical trials are conducted in sequential phases. Phase I trials typically evaluate safety, dose tolerance, and side effects in a small group of participants. Phase II trials explore early indicators of efficacy while continuing to monitor safety in a larger cohort. Phase III trials assess the therapy's overall risk–benefit profile in a broad population, generating the pivotal data required for regulatory approval. Study phases may be split (e.g., phase IIa/IIb) to answer specific research questions or combined (e.g., phase I/II) to accelerate development. At key points, such as the end of phase I or phase II, sponsors often hold formal meetings with the FDA to review study progress, address CMC concerns, and align on future trial design. Clinical trial protocols vary in design and may include randomization, use of control arms, and different levels of blinding depending on the research question. Throughout clinical development, maintaining the QMS ensures standardized documentation, regulatory compliance, and data integrity. cGMP principles must be followed in the manufacturing process, even during early trials, although exemptions for phase I trials may apply in specific contexts [609].

Once sufficient evidence is gathered, the sponsor submits a BLA, which contains comprehensive data on the product's manufacturing, safety, and effectiveness, as well as proposed labeling. Before submission, a pre-BLA meeting is recommended to clarify any outstanding regulatory issues, including statistical methodology and CMC readiness. The FDA has 60 days to assess whether the application is sufficiently complete to file, after which a standard review timeline of 12 months applies. Advisory committee input may be solicited if needed. If approved, the product becomes eligible for marketing in the United States under the terms outlined in the BLA.

6.1.4 Translational Status of Tissue-Engineered Cartilage Products

Despite the publication of over 200 large animal cartilage repair studies over a decade (2013–2022), the number of tissue-engineered cartilage therapies reaching regulatory approval remains limited [232]. In the United States, MACI (Vericel Corp.) remains the only cartilage repair product cleared for marketing under the 351 HCT/P pathway; Carticel, its predecessor, was discontinued after MACI's approval. Globally, different forms of ACI have been approved, typically limited to one approved product per region. Importantly, MACI and other ACI products have narrow indications, primarily for treating focal cartilage defects in the knee, with one exception in the ankle, and remain contraindicated for OA and rheumatoid arthritis (RA). As such, tissue-engineered cartilage products have yet to reach a large portion of the patient population, including those patients requiring treatment in the other cartilage types or anatomical locations.

Clinical outcomes of ACI products have been mixed. A systematic review of MACI with a minimum 10-year follow-up reported durable, long-term improvements in patient-reported outcomes and a low reoperation rate of 9.0% [610]. On the other hand, other recent reviews have indicated failure rates of 14.9%, with most failures occurring within the first five years postoperatively, and failure rates as high as 35.7% for revision ACI surgeries performed after failed primary cartilage procedures [611, 612]. Additionally, multiple studies have shown that the repair tissue formed is predominantly fibrocartilaginous in nature, which is mechanically inferior to the native hyaline cartilage [32, 35, 613]. While MACI provides a precedent for regulatory approval under the 351 HCT/P pathway, its clinical limitations underscore the ongoing need for innovation. In addition, evolving regulatory standards, including more stringent requirements for animal testing, may present further challenges for future products pursuing a similar route.

Despite MACI serving as a regulatory precedent, the translation of tissue-engineered cartilage products to market remains challenging. For example, NeoCart, another form of ACI targeting hyaline cartilage in the knee, pursued approval under the 351 HCT/P pathway but failed to meet its primary endpoint in a phase III clinical trial [231]. Currently, no phase III trials are underway for tissue-engineered cartilage products, highlighting the persistent difficulty of advancing such products toward a BLA submission. Clinical trials in early phases are also limited for tissue-engineered cartilage products. In the United States, CartiLife (Biosolution), a pellet-type matrix-associated autologous chondrocyte product, is undergoing phase II evaluation for the treatment of articular hyaline cartilage of the knee [614]. AuriNovo (3DBio Therapeutics, now PrintBio), a 3D-bioprinted implant containing autologous auricular chondrocytes and collagen hydrogel, was tested in a phase I/IIa clinical trial for microtic ear reconstruction [615]. However, after enrolling only two of the planned eleven patients, the trial was terminated in 2023 because of a company decision unrelated to safety.

Internationally, several noteworthy clinical trials have been conducted. A multicenter phase II trial in Switzerland, Germany, Italy, and Croatia evaluated the use of autologous nasal chondrocytes, either matured or undifferentiated, to treat full-thickness knee cartilage defects [256]. Results showed improved outcomes with matured implants, though no follow-up trials have been reported. In the United Kingdom, a first-in-human trial evaluated undifferentiated MSCs seeded on a collagen scaffold for treating meniscal tears [299]. Although five patients showed clinical improvement, two patients required subsequent meniscectomies due to retear or lack of healing. In Mexico, polyurethane scaffolds enriched with MSCs were tested in 11 patients with partial meniscus defects [616]. The implants maintained adjacent cartilage at 12 months; however, MSC-enriched scaffolds showed no measurable advantage over acellular scaffolds. Together, these early-phase trials offer some promise but also emphasize the difficulty of advancing to pivotal phase III studies required for regulatory approval.

In contrast to traditional tissue engineering approaches, cell-based therapies targeting the IVD, fibrocartilaginous tissue with a nucleus pulposus core, are progressing in the translational pathway. Although not strictly categorized as tissue engineering, these studies are worth highlighting because of their relevance to regenerative strategies for cartilaginous structures. In the United States, DiscGenics' rebonuputemcel injectable disc cell therapy (IDCT) and Mesoblast's rexlemestrocel-L have both received RMAT designations for treating degenerative disc disease. IDCT, composed of allogeneic nucleus pulposus-derived cells in a sodium hyaluronate carrier, has demonstrated improved clinical outcomes and disc volume in phase I/II studies conducted in Japan (38 patients) and the United States (60 patients) [617–619]. Based on these results, a phase III clinical trial has received FDA clearance. Rexlemestrocel-L, an intradiscal injection of allogeneic STRO-3+ BM-MSCs combined with HA, showed significant pain reduction in a phase III trial (404 patients) [620]. A confirmatory phase III trial with approximately 300 patients is now underway [621].

These recent developments illustrate both the promise and the persistent challenges of translating tissue-engineered cartilage technologies. While early-phase studies highlight scientific progress, the absence of late-phase, tissue engineering-based clinical trials for diverse cartilage types indicates the need for continued efforts toward translation.

6.2 Intellectual Property (IP)

IP provides a critical foundation for the commercialization of tissue-engineered cartilage technologies. Because these products often involve complex combinations of cell sources, scaffold materials, delivery systems, and manufacturing techniques, securing robust IP rights is essential for protecting technologies and, perhaps more importantly, attracting investments to fund translational activities. As categorized by the United States Patent and Trademark Office (USPTO), patents generally fall into three types: plant, design, and utility patents. Plant patents apply to certain newly discovered plants; design patents protect the ornamental, non-functional aspects of a product; and utility patents, which make up the vast majority of filings, protect inventions related to processes, machines, compositions of matter, or methods of use. In the context of cartilage tissue engineering, utility patents are most relevant, covering innovations such as chondrocyte isolation or culture methods, scaffold or biomaterial compositions, in vitro or in vivo maturation techniques, and systems for delivering tissue-engineered implants.

In academic settings, the path to securing a patent usually begins with the identification of a novel and potentially translational aspect of the research. When researchers believe their discovery may have translational or clinical relevance, they submit an invention disclosure to their institution's TTO. The TTO then assesses the invention's novelty, patentability, and potential for translation, and may choose to file a provisional patent application to the USPTO to establish an early filing (priority) date. This approach allows researchers to further develop the technology and share findings publicly without immediately jeopardizing future patent rights. However, public disclosure prior to filing can compromise the ability to secure patent protection in many jurisdictions. While some countries, including the United States, Japan, South Korea, China, Mexico, Argentina, and Australia, offer a grace period (commonly 12 months) during which a patent application can still be filed after public disclosure, others, such as those in the European Union, do not. As a result, premature disclosure can limit international patent opportunities. Academic researchers are therefore strongly encouraged to seek guidance on disclosure strategy early in the process to ensure both the timely dissemination of scientific findings and the preservation of IP rights.

Once a provisional patent is filed, researchers typically prepare a non-provisional patent application within 12 months to preserve the priority date. Patent examiners review the submission and may issue objections or rejections, often requiring clarification or amendment. This back-and-forth process, known as patent prosecution, can take several years and usually involves coordination with the TTO or other legal counsel. If granted, the patent becomes a valuable asset that enables licensing to other companies or attracting investments for supporting translational activities.

Academic institutions usually retain ownership of the patents filed under their auspices and may license them to external companies or academic startups founded by the inventors. Licensing agreements often include negotiated terms such as royalties, milestone payments, or equity shares. In some cases, if the university decides not to pursue a patent, inventors may be given the option to file and commercialize the technology independently. Regardless of the pathway, effective management of public disclosure timing, alignment with institutional IP policies, and coordination with legal teams are essential to protect and translate promising tissue engineering innovations.

6.3 Funding and Resource Requirements

Funding and resource needs evolve as a tissue-engineered cartilage product advances along the translational pathway. When academic researchers choose to pursue commercialization themselves as opposed to licensing out their technologies to other companies, they may establish academic startups, which often require funding strategies distinct from those in traditional research settings. Figure 6.2 illustrates a typical progression of funding mechanisms available to academic startups developing tissue-engineered cartilage products. Although individual paths may vary, the process generally begins with early-stage support and advances toward more complex and capital-intensive funding sources, such as VC funding, as the product approaches clinical development and market readiness.

At the outset, non-dilutive funding sources are commonly used to support exploratory research and proof-of-concept studies. These include institutional grants, philanthropic funding, and federal programs such as the Small Business Innovation Research (SBIR) and Small Business Technology Transfer (STTR) grants offered by agencies like the National Institutes of Health (NIH). Such funding mechanisms are particularly valuable

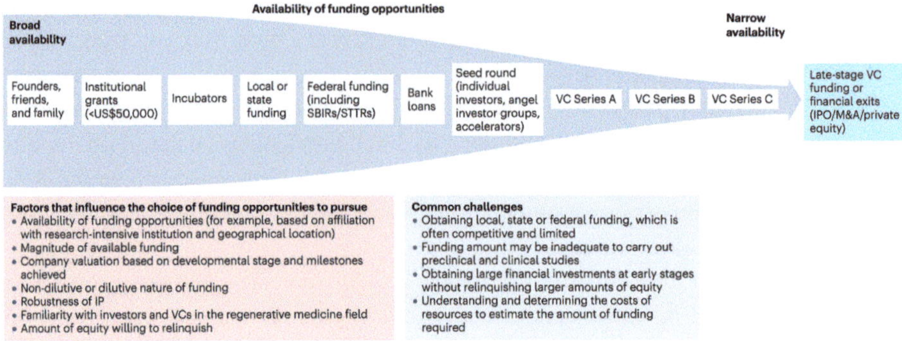

Fig. 6.2 Funding pathways for academic startups. Academic startups may utilize a variety of funding sources throughout the commercialization process. Although the exact sequence varies, many startups begin with institutional grants, incubator support, or initial seed investments before pursuing federal funding opportunities such as Small Business Innovation Research (SBIR) or Small Business Technology Transfer (STTR) grants, as well as regional or accelerator-based funding. Bank loans, while less common, may also serve as an option. As the company grows, it may access venture capital (VC) in the form of Series A, B, or C rounds. Over time, available funding options tend to narrow, and capital needs become more substantial. Mature startups may pursue an initial public offering (IPO), merge with another company, or be acquired by a larger medical company. The type and size of funding pursued often depend on factors such as program availability, whether the funding is dilutive or non-dilutive, and the founders' willingness to give up equity. Key challenges include securing funding in competitive environments and accurately projecting the financial resources needed for development, regulatory approval, and commercialization. M&A, mergers and acquisitions; IP, intellectual property. Reproduced with permission [582]

because they provide capital without requiring founders to relinquish equity, while also serving as early signals of reduced risk for future investors.

As development moves into the preclinical phase, additional resources are required for more structured activities, such as GLP-compliant animal studies including toxicology studies and pivotal large animal studies. A significant amount of funding is necessary to conduct IND-enabling studies; this funding is often dilutive and may come from seed rounds from angel investors or VC firms. Academic institutions may assist with internal translational funds or provide access to incubators that offer laboratory space, operational support, and mentorship. At this stage, collaborations with contract research organizations (CROs) and regulatory consultants become important to ensure that preclinical studies meet FDA expectations and support an eventual IND application submission.

Upon approval of an IND application, the financial demands increase substantially. Even early-phase clinical trials require significant funds to enable patient recruitment, trial site coordination, cGMP-compliant manufacturing, and data infrastructure. To meet these costs, many academic startups seek additional dilutive funding. The success of these fundraising efforts typically hinges on strong IP, a compelling scientific and clinical rationale, and demonstrated progress toward regulatory milestones such as IND clearance.

As the startup matures, it may pursue additional funding to support late-stage clinical trials and additional commercialization activities. These later stages may involve Series A, B, or C rounds of VC funding to support large, multicenter clinical trials and scaling-up of manufacturing. Accelerators, regional economic initiatives, or government-backed loans may also provide supplementary resources, though these become less accessible as capital requirements grow. Eventually, financial exit strategies, such as initial public offerings (IPOs), mergers, or acquisitions, may be used to access further funding, expand market reach, and support large-scale manufacturing and distribution.

Throughout this progression, the availability of funding opportunities tends to narrow, and the magnitude of capital requirements increases. Factors influencing the choice of funding mechanism include the competitiveness of available programs, whether the funding is dilutive or non-dilutive, and the founders' willingness to give up equity. Common challenges include securing adequate funding without excessive dilution and accurately forecasting the costs associated with each development phase. Ultimately, successful translation of a tissue-engineered cartilage therapy depends not only on scientific innovation but also on the strategic alignment of financial resources, infrastructure, and personnel. Aligning funding timelines with developmental milestones, engaging knowledgeable partners, and building a scalable strategy are essential steps in bringing laboratory discoveries to clinical application.

Chapter concepts:

- Successful translation of tissue-engineered cartilage products requires not only scientific innovation but also an understanding of regulatory, IP, and resource requirements.
- Navigating the United States FDA regulatory framework, especially the 351 HCT/P pathway under CBER, is fundamental for the translational success of cell-based therapies.
- Whether a product is classified as a 361 or 351 HCT/P determines the extent of preclinical, clinical, and manufacturing requirements.
- Developers must consult both FDA guidance and international standards such as ISO, ICH, and ASTM to ensure comprehensive regulatory compliance.
- The RMAT designation provides eligible regenerative therapies with early FDA engagement and potential acceleration through the regulatory process.
- Securing robust IP is critical for protecting innovations, attracting investment, and enabling commercialization.
- Funding mechanisms from non-dilutive grants to VC funding must be strategically aligned with developmental milestones and capital demands.
- Market approval remains rare for tissue-engineered cartilage products, but ongoing early-phase trials reflect sustained interest and clinical need.

7 Perspectives

Rachel C. Nordberg, Takumi Takahashi, Jerry C. Hu,
and Kyriacos A. Athanasiou

Cartilage is often thought of as a single tissue type, but, in reality, there are numerous cartilages throughout the body, each endowed with its own unique structure to serve its specialized functions. While articular cartilage of the knee is often the most widely studied cartilage and has received much attention in tissue engineering efforts, there are many other clinically relevant cartilages that also merit attention. In this chapter, we highlight both challenges and areas for future work within the field of cartilage tissue engineering. The specific points described are: (1) conducting additional characterizations of native cartilages, (2) developing joint or tissue-specific assays, (3) using tissue engineering techniques to recapitulate the structure and function of distinct cartilage types (e.g., hyaline vs. fibrocartilage), (4) comparing the engineered implants to native tissues quantitatively, (5) selecting appropriate methods for endpoints and preclinical assessment of implants, (6) translating products from academia to clinical use, and (7) adapting tissue engineering methods to withstand an inflamed environment characteristic of arthritic joints. Overall, we aim to offer the reader a high-level overview of key challenges in the field that impede the clinical translation of tissue-engineered cartilage.

Design standards for an implant can be defined from native tissue characterization studies. As described in Chap. 2, the current understanding of the structure and function of native cartilages is incomplete. Some cartilage tissues are studied far more often than others (e.g., the meniscus is widely studied, but there are very few studies on sacroiliac joint cartilage). Therefore, there are numerous opportunities for additional native tissue characterization studies in both the human and relevant animal models. To exacerbate the incomplete understanding of the structure and function of native cartilages, variability in testing protocols used and reporting standards can make comparing functional values across studies challenging. For example, in the ear, biochemical characterizations at times

R. C. Nordberg et al., *Tissue Engineering the Cartilages of the Body*, Synthesis Lectures on Tissue Engineering, https://doi.org/10.1007/978-3-032-09105-5_7

report hydroxyproline content as a surrogate for collagen content; this makes directly comparing with studies that report collagen content as a percentage of wet or dry weight challenging. Another example is that despite ear cartilage being classified as an elastic cartilage, its quantitative elastin content remains elusive since elastin can be challenging to quantify through methods such as Orbitrap. Mechanical testing results are also highly dependent on the parameters used. For example, testing a sample in uniaxial tension at a strain rate of 10 mm/min may yield different results than testing at 0.5 mm/min. Future studies would benefit from using standardized testing methodologies or testing multiple tissues within the same study to allow for easy comparisons among cartilage types.

While standardized assays to assess the functional properties on the tissue level (e.g., mechanical properties, biochemical content) would enable better cartilage-to-cartilage comparisons to be made, it is also important to consider that different cartilages have different functional requirements and, therefore, joint or location-specific assays may be required. For example, arthroscopy can be performed in large animal knee repair studies [559], but arthroscopy is not generally conducted in other joints such as the TMJ or facet joint. Arthroscopy protocols may need to be adapted or significantly modified to allow for use in such cases. Assays such as gait analysis, pain analysis, or ICRS scoring may also need to be developed to be joint or tissue specific. Additionally, other experimental conditions may need to be tailored to specific applications. For example, non-weight-bearing cartilages (e.g., ear cartilage) may have different mechanical requirements than weight-bearing cartilages (e.g., knee cartilage). Additionally, parameters such as the duration of the pivotal large animal study may need to be adjusted depending on the indication. In FDA guidance for repair of cartilage in the knee, the guidance specifies that a year-long study is generally needed "to provide an adequate period for completion of healing" [538]. It is currently unclear whether this applies to other indications, as formal guidance documents are lacking for cartilage repair in non-knee joints. Therefore, the development of new FDA guidance documents specific to cartilages besides those of the knee would greatly benefit translational efforts within the field.

An implant should be tuned to meet the design standards defined through native tissue characterizations, and these requirements may be substantially different depending on the tissue to be targeted (e.g., hyaline vs. fibrocartilage). As described in Chap. 3, there are numerous potential ways to engineer an implant including both scaffold-based and scaffold-free approaches. As researchers intend to recapitulate different cartilage types, different stimulation regimens may be considered. For example, applying tension may encourage fiber alignment and higher tensile properties [491], which would be particularly useful in engineering structures with a high degree of collagen fiber alignment, such as in the meniscus. Additionally, different starting cell types can be contemplated. For example, the hyaline cartilage to fibrocartilage spectrum can be recapitulated by generating self-assembled neocartilage using chondrocytes that have been rejuvenated in aggregate culture for variable lengths of time [622]. Overall, the tissue engineering regimens should be tuned for the specific cartilage type of interest.

As discussed in Chap. 4, implants should be thoroughly evaluated in vitro before moving on to preclinical studies. Standard tests may include an assessment of gross morphology, histological evaluation of tissue organization, assays to quantify biochemical composition, and biomechanical testing. In addition, advanced imaging techniques or sequencing technologies can be used to augment the understanding of the structure and function of the tissue-engineered cartilage. While some evaluation methods, such as assessing the gross morphology of the implant, may provide straightforward pass/fail decisions, other quantitative criteria may be more complex to compare against native tissue values. An FI can be used to aggregate multiple quantitative parameters into a single value [428, 535, 536]. Overall, because there is no standard guideline for how well neocartilage must mimic the native tissue, it is up to the researcher to decide on what threshold is appropriate for its given application.

After in vitro development has been completed, it is imperative to carefully select an appropriate animal model for preclinical testing. As discussed in Chap. 5, large animal models are preferred for orthopaedic applications due to their ability to recapitulate the human tissue in terms of size and loading conditions. For example, studies that target repair of the TMJ disc have been largely conducted in the Yucatan minipig [235, 556]. Once an animal model is selected, protocols will need to be developed to manufacture an animal cell analog to the product (i.e., ACP). If the product is allogeneic in nature, an animal cell bank may need to be developed [234]. Finally, a study design must be determined for preclinical assessments of the product, which should be finalized in communication with regulatory agencies (e.g., the FDA if in the United States). Researchers should consider including non-terminal endpoints (e.g., in-life assays for safety and function) in conjunction with tissue-level characterizations (e.g., mechanical properties of the repair tissue). Non-terminal endpoints allow for longitudinal tracking of the implant, may reduce the number of animals needed for a study, and are more analogous to clinical endpoints. Overall, careful planning through the preclinical stage will give researchers the best possible chance of advancing toward clinical trials while working within the resource restraints (e.g., funding, time) of an academic environment.

While much of the research on cartilage tissue engineering is conducted at universities [582], the path toward translating novel therapies to clinical translation can be particularly challenging in an academic setting due to a lack of training on the regulatory process, limited funding sources that support translational work, challenges in managing IP issues, and competing demands of an academic career that do not afford much time for translational efforts. While many of these issues are systemic problems that will require changes at a university level, agency level, or the field as a whole, as an individual researcher, being aware of the challenges can allow one to have the best chance of success. Learning about the regulatory process and the steps that need to be taken to translate a product for clinical use can aid researchers in positioning themselves and their research to facilitate translation. For example, regenerative medicine products may pursue RMAT designation to expedite the approval process [555]. Additionally, establishing a preclinical cell bank to manufacture an ACP can streamline the process of conducting preclinical studies [234]. Researchers should also be cognizant of potential IP arising from their work, so as not to disclose data or inventions until appropriate patent applications have been filed. Due to the high costs of conducting preclinical research, researchers may benefit from pursuing non-dilutive funding sources during product development, such as SBIR or STTR grants. For a more thorough discussion of the regulatory and translational challenges that face academic research, please refer to recent reviews on this topic [555, 582].

While this book has primarily focused on the structure and function of healthy native tissues and how to replicate them, a critical point for future cartilage tissue engineering to address is that most of the patients who require cartilage replacements have arthritic degeneration rather than acute injuries. An arthritic joint has an inflamed environment that will likely trigger degenerative changes in the tissue-engineered implants, leading to their degradation in the same manner as the original cartilage. Therefore, methods are needed to modulate inflammation within the repair site, such as through using immunomodulatory biomaterials or synthetic biology approaches, as recently reviewed [231]. It is likely that within the coming decades, the field of "tissue immunoengineering" will receive more emphasis, as next-generation therapies are developed that not only recapitulate the native tissue structure and function but have properties that can modulate the native immune system.

In closing, we hope this book has provided a clear and comprehensive overview of the evolving field of cartilage tissue engineering. The core principles of the tissue engineering process can be distilled into five essential steps: (1) defining design criteria based on the properties of native tissue, (2) engineering neocartilage that recapitulate these properties, (3) rigorously comparing tissue-engineered cartilage to native tissue, (4) assessing safety and efficacy of the implant in relevant animal models, and (5) translating the product toward clinical application. We have structured this narrative to emphasize this process, highlighting both foundational studies and areas of potential future innovation. As this field continues to advance, we encourage the reader to use these concepts and apply them to their own areas of interest to shape the next generation of medical advancements.

References

1. Nasu M, Takayama S, Amagase R, Umezawa A. Effects of basic fibroblast growth factor on cartilage to bone: Time-course histological analysis of in vivo cartilage formation from polydactyly-derived chondrocytes. *Regen Ther* 2025;**28**:462–72. https://doi.org/10.1016/j.reth.2025.01.013.
2. Athanasiou K, Darling E, Hu J, DuRaine G, Reddi H. *Articular Cartilage, Second Edition.* 2017.
3. Kwon H, Brown WE, Lee CA, Wang D, Paschos N, Hu JC, *et al.* Surgical and tissue engineering strategies for articular cartilage and meniscus repair. *Nat Rev Rheumatol* 2019;**15**:550–70. https://doi.org/10.1038/s41584-019-0255-1.
4. Brown WE, Lavernia L, Bielajew BJ, Hu JC, Athanasiou KA. Human nasal cartilage: Functional properties and structure-function relationships for the development of tissue engineering design criteria. *Acta Biomater* 2023;**168**:113–24. https://doi.org/10.1016/j.actbio.2023.07.011.
5. Huwe LW, Brown WE, Hu JC, Athanasiou KA. Characterization of costal cartilage and its suitability as a cell source for articular cartilage tissue engineering. *J Tissue Eng Regen Med* 2018;**12**:1163–76. https://doi.org/10.1002/term.2630.
6. Donahue RP, Gonzalez-Leon EA, Hu JC, Athanasiou KA. Considerations for translation of tissue engineered fibrocartilage from bench to bedside. *J Biomech Eng* 2018;**141**:. https://doi.org/10.1115/1.4042201.
7. Kobayashia S, Takebe T, Inui M, Iwai S, Kan H, Zheng YW, *et al.* Reconstruction of human elastic cartilage by a CD44 + CD90 + stem cell in the ear perichondrium. *Proc Natl Acad Sci U S A* 2011;**108**:14479–84. https://doi.org/10.1073/pnas.1109767108.
8. Gonzales N, Garrity C, Rivas I, McEligot H, Vapniarsky N. Auricular Chondrocytes as a Cell Source for Scaffold-Free Elastic Cartilage Tissue Engineering. *Tissue Eng Part C Methods* 2024;**30**:314–22. https://doi.org/10.1089/ten.tec.2024.0106.
9. Ramos R, Pham KT, Prince RC, Leiser-Miller LB, Prasad MS, Wang X, *et al.* Superstable lipid vacuoles endow cartilage with its shape and biomechanics. *Science (1979)* 2025;**387**:eads9960. https://doi.org/10.1126/science.ads9960.
10. Betts JG, Young KA, Wise JA, Johnson E, Poe B, Kruse DH, *et al. Anatomy and Physiology.* Houston, Texas: OpenStax; 2013.
11. Bielajew BJ, Hu JC, Athanasiou KA. Collagen: quantification, biomechanics and role of minor subtypes in cartilage. *Nat Rev Mater* 2020;**5**:730–47. https://doi.org/10.1038/s41578-020-0213-1.
12. Kim KW, Wong ME, Helfrick JF, Thomas JB, Athansiou KA. Biomechanical Tissue Characterization of the Superior Joint Space of the Porcine Temporomandibular Joint. *Annals of Biomedical Engineering 2003 31:8* 2003;**31**:924–30. https://doi.org/10.1114/1.1591190.

R. C. Nordberg et al., *Tissue Engineering the Cartilages of the Body*, Synthesis Lectures on Tissue Engineering, https://doi.org/10.1007/978-3-032-09105-5

13. Ichinose S, Yamagata K, Sekiya I, Muneta T, Tagami M. Detailed examination of cartilage formation and endochondral ossification using human mesenchymal stem cells. *Clin Exp Pharmacol Physiol* 2005;**32**:561–70. https://doi.org/10.1111/j.1440-1681.2005.04231.x.
14. Carter DR, Beaupré GS, Wong M, Smith RL, Andriacchi TP, Schurman DJ. The Mechanobiology of Articular Cartilage Development and Degeneration.
15. Fox AJS, Sophia Fox AJ, Bedi A, Rodeo SA. The Basic Science of Articular Cartilage: Structure, Composition, and Function. *Sports Health: A Multidisciplinary Approach* 2009;**1**:461–8. https://doi.org/10.1177/1941738109350438.
16. Roughley PJ, Mort JS. The role of aggrecan in normal and osteoarthritic cartilage. *J Exp Orthop* 2014;**1**:8. https://doi.org/10.1186/s40634-014-0008-7.
17. Donnan FG. Theorie der Membrangleichgewichte und Membranpotentiale bei Vorhandensein von nicht dialysierenden Elektrolyten. Ein Beitrag zur physikalisch-chemischen Physiologie. *Zeitschrift Für Elektrochemie Und Angewandte Physikalische Chemie /* 1911;**17**:572–81. https://doi.org/10.1002/bbpc.19110171405.
18. Flowers SA, Zieba A, Örnros J, Jin C, Rolfson O, Björkman LI, *et al.* Lubricin binds cartilage proteins, cartilage oligomeric matrix protein, fibronectin and collagen II at the cartilage surface. *Sci Rep* 2017;**7**:. https://doi.org/10.1038/s41598-017-13558-y.
19. Bhosale AM, Richardson JB. Articular cartilage: structure, injuries and review of management. *Br Med Bull* 2008;**87**:77–95. https://doi.org/10.1093/bmb/ldn025.
20. Van Mourik M, Schuiringa GH, Varion-Verhagen LP, Vonk LA, Van Donkelaar CC, Ito K, *et al.* Enzymatic Isolation of Articular Chondrons: Is It Much Different Than That of Chondrocytes? *Tissue Eng Part C Methods* 2023;**29**:30–40. https://doi.org/10.1089/ten.tec.2022.0176.
21. Alexopoulos LG, Setton LA, Guilak F. The biomechanical role of the chondrocyte pericellular matrix in articular cartilage. *Acta Biomater* 2005;**1**:317–25. https://doi.org/10.1016/j.actbio.2005.02.001.
22. Poole CA, Flint MH, Beaumont BW. Chondrons in cartilage: Ultrastructural analysis of the pericellular microenvironment in adult human articular cartilages. *Journal of Orthopaedic Research* 1987;**5**:509–22. https://doi.org/10.1002/jor.1100050406.
23. Peng G, McNary SM, Athanasiou KA, Reddi AH. Superficial Zone Extracellular Matrix Extracts Enhance Boundary Lubrication of Self-Assembled Articular Cartilage. *Cartilage* 2016;**7**:256–64. https://doi.org/10.1177/1947603515612190.
24. Mow VC, Ratcliffe A, Poole AR. *REVIEW Cartilage and diarthrodial joints as paradigms for hierarchical materials and structures.* n.d.
25. Peng G, McNary SM, Athanasiou KA, Reddi AH. The distribution of superficial zone protein (SZP)/lubricin/PRG4 and boundary mode frictional properties of the bovine diarthrodial joint. *J Biomech* 2015;**48**:3406–12. https://doi.org/10.1016/j.jbiomech.2015.05.032.
26. Iqbal SM, Leonard C, Regmi SC, De Rantere D, Tailor P, Ren G, *et al.* Lubricin/Proteoglycan 4 binds to and regulates the activity of Toll-Like Receptors in Vitro. *Sci Rep* 2016;**6**:. https://doi.org/10.1038/srep18910.
27. Kwiecinski JJ, Dorosz SG, Ludwig TE, Abubacker S, Cowman MK, Schmidt TA. The effect of molecular weight on hyaluronan's cartilage boundary lubricating ability - alone and in combination with proteoglycan 4. *Osteoarthritis Cartilage* 2011;**19**:1356–62. https://doi.org/10.1016/j.joca.2011.07.019.
28. Eschweiler J, Horn N, Rath B, Betsch M, Baroncini A, Tingart M, *et al.* The biomechanics of cartilage-an overview. *Life* 2021;**11**:. https://doi.org/10.3390/life11040302.
29. Huang Y, Fan H, Gong X, Yang L, Wang F. Scaffold With Natural Calcified Cartilage Zone for Osteochondral Defect Repair in Minipigs. *American Journal of Sports Medicine* 2021;**49**:1883–91. https://doi.org/10.1177/03635465211007139.

30. Kobayashi S, Takebe T, Zheng YW, Mizuno M, Yabuki Y, Maegawa J, *et al.* Presence of cartilage stem/progenitor cells in adult mice auricular perichondrium. *PLoS One* 2011;**6**:. https://doi.org/10.1371/journal.pone.0026393.
31. Salinas EY, Otarola GA, Kwon H, Wang D, Hu JC, Athanasiou KA, *et al.* Topographical characterization of the young, healthy human femoral medial condyle. *Cartilage* 2023;**14**:338–50. https://doi.org/10.1177/19476035221141421.
32. Alizadeh Sardroud H, Wanlin T, Chen X, Eames BF. Cartilage Tissue Engineering Approaches Need to Assess Fibrocartilage When Hydrogel Constructs Are Mechanically Loaded. *Front Bioeng Biotechnol* 2022. https://doi.org/10.3389/fbioe.2021.787538.
33. Link JM, Salinas EY, Hu JC, Athanasiou KA. The tribology of cartilage: Mechanisms, experimental techniques, and relevance to translational tissue engineering. *Clinical Biomechanics* 2019. https://doi.org/10.1016/j.clinbiomech.2019.10.016.
34. Akkiraju H, Nohe A. Role of Chondrocytes in Cartilage Formation, Progression of Osteoarthritis and Cartilage Regeneration. *J Dev Biol* 2015;**3**:177–92. https://doi.org/10.3390/jdb3040177.
35. Armiento AR, Alini M, Stoddart MJ. Articular fibrocartilage - Why does hyaline cartilage fail to repair? *Adv Drug Deliv Rev* 2019;**146**:289–305. https://doi.org/10.1016/j.addr.2018.12.015.
36. Schwartz AG, Long F, Thomopoulos S. Enthesis fibrocartilage cells originate from a population of hedgehog-responsive cells modulated by the loading environment. *Development (Cambridge)* 2015;**142**:196–206. https://doi.org/10.1242/dev.112714.
37. Gkantsinikoudis N, Kapetanakis S, Magras I, Tsiridis E, Kritis A. Tissue Engineering of Human Intervertebral Disc: A Concise Review. *Tissue Eng Part B Rev* 2022:848–60. https://doi.org/10.1089/ten.teb.2021.0090.
38. Athanasiou KA, Almarza AA, Detamore MS, Kalpakci KN. *Tissue Engineering of Temporomandibular Joint Cartilage.* Cham: Springer International Publishing; 2009.
39. Kalpakci KN, Kim EJ, Athanasiou KA. Assessment of growth factor treatment on fibrochondrocyte and chondrocyte co-cultures for TMJ fibrocartilage engineering. *Acta Biomater* 2011;**7**:1710–8. https://doi.org/10.1016/j.actbio.2010.12.015.
40. Seon J. Biomechanical Properties of Fibrocartilage. In: Koh J, Zaffagnini S, Kuroda R, Longo UG, Amirouche F, editors. *Orthopaedic Biomechanics in Sports Medicine.* Springer International Publishing; 2021. p. 73–9.
41. Athanasiou KA, Sanchez-Adams J. *Engineering the Knee Meniscus.* Cham: Springer International Publishing; 2009.
42. Chan SCW, Ferguson SJ, Gantenbein-Ritter B. The effects of dynamic loading on the intervertebral disc. *European Spine Journal* 2011:1796–812. https://doi.org/10.1007/s00586-011-1827-1.
43. Gonzalez-Leon EA, Hu JC, Athanasiou KA. Yucatan Minipig Knee Meniscus Regional Biomechanics and Biochemical Structure Support its Suitability as a Large Animal Model for Translational Research. *Front Bioeng Biotechnol* 2022;**10**:. https://doi.org/10.3389/FBIOE.2022.844416.
44. Lyu FJ, Cheung KM, Zheng Z, Wang H, Sakai D, Leung VY. IVD progenitor cells: a new horizon for understanding disc homeostasis and repair. *Nat Rev Rheumatol* 2019:102–12. https://doi.org/10.1038/s41584-018-0154-x.
45. Aharonov A, Sofer S, Bruck H, Sarig U, Sharabi M. Unveiling the mechanical role of radial fibers in meniscal tissue: Toward structural biomimetics. *Acta Biomater* 2024;**187**:199–211. https://doi.org/10.1016/j.actbio.2024.08.024.
46. Becker I, Woodley SJ, Stringer MD. The adult human pubic symphysis: A systematic review. *J Anat* 2010:475–87. https://doi.org/10.1111/j.1469-7580.2010.01300.x.

47. Nguyen MS, Kheyfits V, Giordano BD, Dieudonne G, Monu JUV. Hip anatomic variants that may mimic abnormalities at MRI: Labral variants. *American Journal of Roentgenology* 2013. https://doi.org/10.2214/AJR.12.9860.
48. Almajed YA, Hall AC, Gillingwater TH, Alashkham A. Anatomical, functional and biomechanical review of the glenoid labrum. *J Anat* 2022:761–71. https://doi.org/10.1111/joa.13582.
49. Riedler KL, Shokrani A, Markarian A, Fisher LM, Pepper JP. Age-related histologic and biochemical changes in auricular and septal cartilage. *Laryngoscope* 2017;**127**:E399–407. https://doi.org/10.1002/lary.26807.
50. Ito I, Imada M, Ikeda M, Sueno K, Arikuni T, Kida A. A morphological study of age changes in adult human auricular cartilage with special emphasis on elastic fibers. *Laryngoscope* 2001;**111**:881–6. https://doi.org/10.1097/00005537-200105000-00023.
51. Posniak S, Chung JHY, Liu X, Mukherjee P, Wallace GG. The importance of elastin and its role in auricular cartilage tissue engineering. *Bioprinting* 2023. https://doi.org/10.1016/j.bprint.2023.e00276.
52. Hill JR, Eekhoff JD, Brophy RH, Lake SP. Elastic fibers in orthopedics: Form and function in tendons and ligaments, clinical implications, and future directions. *Journal of Orthopaedic Research* 2020:2305–17. https://doi.org/10.1002/jor.24695.
53. Gvaramia D, Kern J, Jakob Y, Zenobi-Wong M, Rotter N. Regenerative Potential of Perichondrium: A Tissue Engineering Perspective. *Tissue Eng Part B Rev* 2021;**28**:531–41. https://doi.org/10.1089/ten.teb.2021.0054.
54. Nam S, Cho W, Cho H, Lee J, Lee E, Son Y. Xiphoid Process-Derived Chondrocytes: A Novel Cell Source for Elastic Cartilage Regeneration. *Stem Cells Transl Med* 2014;**3**:1381–91. https://doi.org/10.5966/sctm.2014-0070.
55. Sun H, Zhou J, Wang Q, Jiang H, Yang Q. Contribution of perichondrium to the mechanical properties of auricular cartilage. *J Biomech* 2021;**126**:. https://doi.org/10.1016/j.jbiomech.2021.110638.
56. Ghate SK, Kalambe A, Maldhure S. Auricular haematoma an avoidable cosmetic deformity: A chance or negligence. *American Journal of Otolaryngology - Head and Neck Medicine and Surgery* 2022;**43**:. https://doi.org/10.1016/j.amjoto.2021.103232.
57. Pauken CM, Heyes R, Lott DG. Mechanical, Cellular, and Proteomic Properties of Laryngotracheal Cartilage. *Cartilage* 2019:321–8. https://doi.org/10.1177/1947603517749921.
58. Mallinger R, Bock P. Differentiation of extracellular matrix in the cellular cartilage of the mouse pinna. *Anat Embryol (Berl)* 1985;**172**:69–74. https://doi.org/10.1007/BF00318945.
59. Švajger A. Chondrogenesis in the external ear of the rat. *Z Anat Entwicklungsgesch* 1970;**131**:236–42. https://doi.org/10.1007/BF00520966.
60. Anderson DR. The ultrastructure of elastic and hyaline cartilage of the rat. *American Journal of Anatomy* 1964;**114**:403–33. https://doi.org/10.1002/aja.1001140305.
61. Sanzone CF, Reith EJ. The development of the elastic cartilage of the mouse pinna. *American Journal of Anatomy* 1976;**146**:31–71. https://doi.org/10.1002/aja.1001460103.
62. Couly GF, Coltey PM, Douarin NM Le. The triple origin of skull in higher vertebrates: a study in quail-chick chimeras. *Development* 1993;**117**:409–29. https://doi.org/10.1242/dev.117.2.409.
63. Olivera-Martinez I, Coltey M, Dhouailly D, Pourquié O. Mediolateral somitic origin of ribs and dermis determined by quail-chick chimeras. *Development* 2000;**127**:4611–7. https://doi.org/10.1242/dev.127.21.4611.
64. Tickle C. The contribution of chicken embryology to the understanding of vertebrate limb development. *Mech Dev* 2004:1019–29. https://doi.org/10.1016/j.mod.2004.05.015.
65. Teillet MA, Ziller C, Le Douarin NM. Quail—Chick Chimeras. In: Sharpe PT, Mason I, editors. *Molecular Embryology Second Edition*. n.d.

66. Davidson ENB, van de Loo FAJ, van den Berg WB, van der Kraan PM. How to build an inducible cartilage-specific transgenic mouse. *Arthritis Res Ther* 2014. https://doi.org/10.1186/ar4573.
67. Pest MA, Beier F. Developmental biology: Is there such a thing as a cartilage-specific knockout mouse? *Nat Rev Rheumatol* 2014:702–4. https://doi.org/10.1038/nrrheum.2014.168.
68. Mizoguchi T. In vivo dynamics of hard tissue-forming cell origins: Insights from Cre/loxP-based cell lineage tracing studies. *Japanese Dental Science Review* 2024:109–19. https://doi.org/10.1016/j.jdsr.2024.01.003.
69. Yelick Pamela C, Schilling Thomas F. Molecular Dissection of Craniofacial Development Using Zebrafish. *Critical Reviews in Oral Biology & Medicine* 2002;**13**:308–22. https://doi.org/10.1177/154411130201300402.
70. Keenan SR, Beck CW. Xenopus Limb Bud Morphogenesis n.d. https://doi.org/10.1002/dvdy.
71. Busse B, Galloway JL, Gray RS, Harris MP, Kwon RY. Zebrafish: An Emerging Model for Orthopedic Research. *Journal of Orthopaedic Research* 2020:925–36. https://doi.org/10.1002/jor.24539.
72. Alibardi L. Appendage regeneration in anamniotes utilizes genes active during larval-metamorphic stages that have been lost or altered in amniotes: The case for studying lizard tail regeneration. *J Morphol* 2020:1358–81. https://doi.org/10.1002/jmor.21251.
73. Prasad MS, Charney RM, García-Castro MI. Specification and formation of the neural crest: Perspectives on lineage segregation. *Genesis* 2019. https://doi.org/10.1002/dvg.23276.
74. Tani S, Chung U il, Ohba S, Hojo H. Understanding paraxial mesoderm development and sclerotome specification for skeletal repair. *Exp Mol Med* 2020:1166–77. https://doi.org/10.1038/s12276-020-0482-1.
75. Salva JE, Merrill AE. Signaling Networks in Joint Development n.d. https://doi.org/10.1002/dvdy.
76. Minoux M, Rijli FM. Molecular mechanisms of cranial neural crest cell migration and patterning in craniofacial development. *Development* 2010:2605–21. https://doi.org/10.1242/dev.040048.
77. Hinton RJ. Genes that Regulate Morphogenesis and Growth of the Temporomandibular Joint: A Review n.d. https://doi.org/10.1002/dvdy.
78. Wyngaarden LA, Vogeli KM, Ciruna BG, Wells M, Hadjantonakis AK, Hopyan S. Oriented cell motility and division underlie early limb bud morphogenesis. *Development* 2010;**137**:2551–8. https://doi.org/10.1242/dev.046987.
79. Tuan RS. *Cellular Signaling in Developmental Chondrogenesis: N-Cadherin, Wnts, and BMP-2.* 2003.
80. Chijimatsu R, Saito T. Mechanisms of synovial joint and articular cartilage development. *Cellular and Molecular Life Sciences* 2019:3939–52. https://doi.org/10.1007/s00018-019-03191-5.
81. Hall BK, Miyake T. All for one and one for all: Condensations and the initiation of skeletal development. *BioEssays* 2000:138–47. https://doi.org/10.1002/(SICI)1521-1878(200002)22:2<138::AID-BIES5>3.0.CO;2-4.
82. Decker RS, Koyama E, Pacifici M. Genesis and morphogenesis of limb synovial joints and articular cartilage. *Matrix Biology* 2014:5–10. https://doi.org/10.1016/j.matbio.2014.08.006.
83. Koyama E, Shibukawa Y, Nagayama M, Sugito H, Young B, Yuasa T, *et al.* A distinct cohort of progenitor cells participates in synovial joint and articular cartilage formation during mouse limb skeletogenesis. *Dev Biol* 2008;**316**:62–73. https://doi.org/10.1016/j.ydbio.2008.01.012.
84. Pacifici M, Koyama E, Iwamoto M. Mechanisms of synovial joint and articular cartilage formation: Recent advances, but many lingering mysteries. *Birth Defects Res C Embryo Today* 2005:237–48. https://doi.org/10.1002/bdrc.20050.

85. Lee SY, Abel ED, Long F. Glucose metabolism induced by Bmp signaling is essential for murine skeletal development. *Nat Commun* 2018;**9**:. https://doi.org/10.1038/s41467-018-07316-5.
86. Blumer MJF, Longato S, Fritsch H. Structure, formation and role of cartilage canals in the developing bone. *Annals of Anatomy* 2008;**190**:305–15. https://doi.org/10.1016/j.aanat.2008.02.004.
87. Guan Y, Yang X, Yang W, Charbonneau C, Chen Q. Mechanical activation of mammalian target of rapamycin pathway is required for cartilage development. *FASEB Journal* 2014;**28**:4470–81. https://doi.org/10.1096/fj.14-252783.
88. Garcia J, Mennan C, McCarthy HS, Roberts S, Richardson JB, Wright KT. Chondrogenic Potency Analyses of Donor-Matched Chondrocytes and Mesenchymal Stem Cells Derived from Bone Marrow, Infrapatellar Fat Pad, and Subcutaneous Fat. *Stem Cells Int* 2016;**2016**:. https://doi.org/10.1155/2016/6969726.
89. Las Heras F, Gahunia HK, Pritzker KPH. Articular Cartilage Development: A Molecular Perspective. *Orthopedic Clinics of North America* 2012;**43**:155–71. https://doi.org/10.1016/j.ocl.2012.01.003.
90. Rhee DK, Marcelino J, Baker M, Gong Y, Smits P, Lefebvre V, *et al.* The secreted glycoprotein lubricin protects cartilage surfaces and inhibits synovial cell overgrowth. *Journal of Clinical Investigation* 2005;**115**:622–31. https://doi.org/10.1172/JCI200522263.
91. Gannon AR, Nagel T, Bell AP, Avery NC, Kelly DJ. The changing role of the superficial region in determining the dynamic compressive properties of articular cartilage during postnatal development. *Osteoarthritis Cartilage* 2015;**23**:975–84. https://doi.org/10.1016/j.joca.2015.02.003.
92. Hunziker EB. Mechanism of longitudinal bone growth and its regulation by growth plate chondrocytes. *Microsc Res Tech* 1994;**28**:505–19. https://doi.org/10.1002/jemt.1070280606.
93. Kronenberg HM. Developmental regulation of the growth plate. *Nature* 2003;**423**:332–6. https://doi.org/10.1038/nature01657.
94. Shea CA, Rolfe RA, Murphy P, Rolfe v R A, Researcher P. The importance of foetal movement for co-ordinated cartilage and bone development in utero CLINICAL CONSEQUENCES AND POTENTIAL FOR THERAPY. *Bone Joint Res* 2015;**4**:105–21. https://doi.org/10.1302/2046-3758.47.
95. Nowlan NC, Murphy P, Prendergast PJ. A dynamic pattern of mechanical stimulation promotes ossification in avian embryonic long bones. *J Biomech* 2008;**41**:249–58. https://doi.org/10.1016/j.jbiomech.2007.09.031.
96. Nowlan NC, Bourdon C, Dumas G, Tajbakhsh S, Prendergast PJ, Murphy P. Developing bones are differentially affected by compromised skeletal muscle formation. *Bone* 2010;**46**:1275–85. https://doi.org/10.1016/j.bone.2009.11.026.
97. Brama PAJ, Tekoppele JM, Bank RA, Barneveld A, Firth EC, Van Weeren PR. The influence of strenuous exercise on collagen characteristics of articular cartilage in Thoroughbreds age 2 years. *Equine Vet J* 2000;**32**:551–4. https://doi.org/10.2746/042516400777584613.
98. Brama PAJ, Tekoppele JM, Bank RA, Barneveld A, Van Weeren PR. Functional adaptation of equine articular cartilage: The formation of regional biochemical characteristics up to age one year. *Equine Vet J* 2000;**32**:217–21. https://doi.org/10.2746/042516400776563626.
99. Koo S, Andriacchi TP. A comparison of the influence of global functional loads vs. local contact anatomy on articular cartilage thickness at the knee. *J Biomech* 2007;**40**:2961–6. https://doi.org/10.1016/j.jbiomech.2007.02.005.
100. Vanwanseele B, Lucchinetti E, Stüssi E. The effects of immobilization on the characteristics of articular cartilage: Current concepts and future directions. *Osteoarthritis Cartilage* 2002;**10**:408–19. https://doi.org/10.1053/joca.2002.0529.

101. Schipani E, Ryan HE, Didrickson S, Kobayashi T, Knight M, Johnson RS. Hypoxia in cartilage: HIF-1α is essential for chondrocyte growth arrest and survival. *Genes Dev* 2001;**15**:2865–76. https://doi.org/10.1101/gad.934301.
102. Robins JC, Akeno N, Mukherjee A, Dalal RR, Aronow BJ, Koopman P, *et al.* Hypoxia induces chondrocyte-specific gene expression in mesenchymal cells in association with transcriptional activation of Sox9. *Bone* 2005;**37**:313–22. https://doi.org/10.1016/j.bone.2005.04.040.
103. Provot S, Schipani E. Fetal Growth Plate: A Developmental Model of Cellular Adaptation to Hypoxia.
104. Hirao M, Tamai N, Tsumaki N, Yoshikawa H, Myoui A. Oxygen tension regulates chondrocyte differentiation and function during endochondral ossification. *J Biol Chem* 2006;**281**:31079–92. M602296200 [pii].
105. Martin JA, Ellerbroek SM, Buckwalter JA. Age-related decline in chondrocyte response to insulin-like growth factor-I: The role of growth factor binding proteins. *Journal of Orthopaedic Research* 1997;**15**:491–8. https://doi.org/10.1002/jor.1100150403.
106. Jørgensen AEM, Kjær M, Heinemeier KM. The effect of aging and mechanical loading on the metabolism of articular cartilage. *Journal of Rheumatology* 2017:410–7. https://doi.org/10.3899/jrheum.160226.
107. Wilson B, Novakofski KD, Donocoff RS, Liang YXA, Fortier LA. Telomerase Activity in Articular Chondrocytes Is Lost after Puberty. *Cartilage* 2014;**5**:215–20. https://doi.org/10.1177/1947603514537518.
108. Verzijl N, Degroot J, Zaken C Ben, Braun-Benjamin O, Maroudas A, Bank RA, *et al.* Crosslinking by Advanced Glycation End Products Increases the Stiffness of the Collagen Network in Human Articular Cartilage A Possible Mechanism Through Which Age Is a Risk Factor for Osteoarthritis. *Arthritis Rheum* 2002;**46**:114–23. https://doi.org/10.1002/art.10025.
109. Wilson SL, Guilbert M, Sulé-Suso J, Torbet J, Jeannesson P, Sockalingum GD, *et al.* A microscopic and macroscopic study of aging collagen on its molecular structure, mechanical properties, and cellular response. *FASEB Journal* 2014;**28**:14–25. https://doi.org/10.1096/fj.13-227579.
110. Dudhia J. Aggrecan, aging and assembly in articular cartilage. *Cellular and Molecular Life Sciences* 2005:2241–56. https://doi.org/10.1007/s00018-005-5217-x.
111. Hjertquist SO, Lemperg R. Identification and concentration of the glycosaminoglycans of human articular cartilage in relation to age and osteoarthritis. *Calc Tis Res* 1972;**10**:223–37.
112. Pearce RH, Mathieson JM, Mort JS, Roughley PJ. Effect of age on the abundance and fragmentation of link protein of the human intervertebral disc. *Journal of Orthopaedic Research* 1989;**7**:861–7. https://doi.org/10.1002/jor.1100070612.
113. Zhang M, Theleman JL, Lygrisse KA, Wang J. Epigenetic mechanisms underlying the aging of articular cartilage and osteoarthritis. *Gerontology* 2019:387–96. https://doi.org/10.1159/000496688.
114. Ray DrK, Bardhan DrJ, Sarkar DrKN. A Study of Calcification of Costal Cartilages (1st To 7th) In Different Age Groups and Its Effect on Chest Expansion in Both Male and Female. *IOSR Journal of Dental and Medical Sciences* 2017;**16**:115–23. https://doi.org/10.9790/0853-160307115123.
115. Donahue RP, Hu JC, Athanasiou KA. Remaining Hurdles for Tissue-Engineering the Temporomandibular Joint Disc. *Trends Mol Med* 2019;**25**:241–56. https://doi.org/10.1016/j.molmed.2018.12.007.
116. Turner AS, Athanasiou~' KA, Zhut C-F, Alvis MR, Bryant HU. Biochemical effects of estrogen on articular cartilage in ovariectomized sheep. *Osteoarthritis Cartilage* 1997;**5**:63–9.

117. King KB, Rosenthal AK. The adverse effects of diabetes on osteoarthritis: update on clinical evidence and molecular mechanisms. *Osteoarthritis Cartilage* 2015;**23**:841–50. https://doi.org/10.1016/j.joca.2015.03.031.
118. Wernecke C, Braun HJ, Dragoo JL. The effect of intra-articular corticosteroids on articular cartilage: A systematic review. *Orthop J Sports Med* 2015;**3**:1–7. https://doi.org/10.1177/2325967115581163.
119. Widuchowski W, Widuchowski J, Trzaska T. Articular cartilage defects: Study of 25,124 knee arthroscopies. *Knee* 2007;**14**:177–82. https://doi.org/10.1016/j.knee.2007.02.001.
120. Radin EL, Martin RB, Burr DB, Caterson B, Boyd RD, Goodwin C. Effects of mechanical loading on the tissues of the rabbit knee. *Journal of Orthopaedic Research* 1984;**2**:221–34. https://doi.org/10.1002/jor.1100020303.
121. Dekel S, Weissman SL. Joint Changes after Overuse and Peak Overloading of Rabbit Knees In Vivo. *Acta Orthop* 1978;**49**:519–28. https://doi.org/10.3109/17453677808993232.
122. Goldring MB, Goldring SR. Articular Cartilage and Subchondral Bone in the Pathogenesis of Osteoarthritis.
123. Buckwalter JA, Mow VC, Ratcliffe A. Restoration of Injured or Degenerated Articular Cartilage. *JAAOS - Journal of the American Academy of Orthopaedic Surgeons* 1994;**2**:.
124. Jin X, Liang W, Zhang L, Cao S, Yang L, Xie F. Economic and Humanistic Burden of Osteoarthritis: An Updated Systematic Review of Large Sample Studies. *Pharmacoeconomics* 2023:1453–67. https://doi.org/10.1007/s40273-023-01296-1.
125. Leifer VP, Katz JN, Losina E. The burden of OA-health services and economics. *Osteoarthritis Cartilage* 2022;**30**:10–6. https://doi.org/10.1016/j.joca.2021.05.007.
126. Ferrari AJ, Santomauro DF, Aali A, Abate YH, Abbafati C, Abbastabar H, *et al.* Global incidence, prevalence, years lived with disability (YLDs), disability-adjusted life-years (DALYs), and healthy life expectancy (HALE) for 371 diseases and injuries in 204 countries and territories and 811 subnational locations, 1990–2021: a systematic analysis for the Global Burden of Disease Study 2021. *The Lancet* 2024;**403**:2133–61. https://doi.org/10.1016/S0140-6736(24)00757-8.
127. Gold M, Boyack I, Caputo N, Pearlman A. Imaging prevalence of nasal septal perforation in an urban population. *Clin Imaging* 2017;**43**:80–2. https://doi.org/10.1016/j.clinimag.2017.02.002.
128. Belmont PJ, McCriskin BJ, Sieg RN, Burks R, Schoenfeld AJ. Combat wounds in Iraq and Afghanistan from 2005 to 2009. *Journal of Trauma and Acute Care Surgery* 2012;**73**:3–12. https://doi.org/10.1097/TA.0b013e318250bfb4.
129. Roblin DG, Eccles R. What, if any, is the value of septal surgery? *Clin Otolaryngol Allied Sci* 2002:77–80. https://doi.org/10.1046/j.1365-2273.2002.00531.x.
130. Englund M, Guermazi A, Gale D, Hunter DJ, Aliabadi P, Clancy M, *et al. Incidental Meniscal Findings on Knee MRI in Middle-Aged and Elderly Persons A bs tr ac t*. vol. 359. 2008.
131. Luvsannyam E, Jain MS, Leitao AR, Maikawa N, Leitao AE. Meniscus Tear: Pathology, Incidence, and Management. *Cureus* 2022. https://doi.org/10.7759/cureus.25121.
132. Xu S, Chen J, Wang C, Lin Y, Huang W, Zhou H, *et al.* Global, regional, and national burden of low back pain for adults aged 55 and older 1990–2021: an analysis for the global burden of disease study 2021. *BMC Musculoskelet Disord* 2025;**26**:81. https://doi.org/10.1186/s12891-025-08326-3.
133. Donahue RP, Kallins EG, Hu JC, Athanasiou KA. Characterization of the Temporomandibular Joint Disc Complex in the Yucatan Minipig. *Tissue Eng Part A* 2023;**29**:439–48. https://doi.org/10.1089/ten.tea.2023.0011.
134. Liu XM, Zhang SY, Yang C, Chen MJ, Cai XY, Haddad MS, *et al.* Correlation between disc displacements and locations of disc perforation in the temporomandibular joint. *Dentomaxillofacial Radiology* 2010;**39**:149–56. https://doi.org/10.1259/dmfr/72395946.

135. Sequeira W. *Diseases of the Pubic Symphysis.* n.d.
136. Stover MD, Edelstein AI, Matta JM. Chronic Anterior Pelvic Instability: Diagnosis and Management. *Journal of the American Academy of Orthopaedic Surgeons* 2017:509–17. https://doi.org/10.5435/JAAOS-D-15-00338.
137. Nitsch A, Marx H, Fischer CS, Bakir S, Ekkernkamp A, Wassilew GI, *et al.* Prevalence of cauliflower ear in high level judoka. *Sci Rep* 2023;**13**:. https://doi.org/10.1038/s41598-023-42635-8.
138. Batur A, Sakarya ME, Uslu A. Osteophyte arising from the cricoid cartilage causing dysphagia. *European Journal of General Medicine* 2014;**11**:45–7. https://doi.org/10.15197/sabad.1.11.11.
139. Takasaki K, Takahashi H, Miyamoto I, Yoshida H, Yamamoto-Fukuda T, Enatsu K, *et al.* Measurement of angle and length of the eustachian tube on computed tomography using the multiplanar reconstruction technique. *Laryngoscope* 2007;**117**:1251–4. https://doi.org/10.1097/MLG.0b013e318058a09f.
140. Athanasiou KA, Agarwal A, Muffoletto A, Dzida FJ, Constantinides G, Clem M. Biomechanical properties of hip cartilage in experimental animal models. *Clin Orthop Relat Res* 1995:254–66.
141. Curl WW, Krome J, Gordon ES, Rushing J, Stat M, Smith BP, *et al. Cartilage Injuries: A Review of 31,516 Knee Arthroscopies.* 1997.
142. Deshpande BR, Katz JN, Solomon DH, Yelin EH, Hunter DJ, Messier SP, *et al.* Number of Persons With Symptomatic Knee Osteoarthritis in the US: Impact of Race and Ethnicity, Age, Sex, and Obesity. *Arthritis Care Res (Hoboken)* 2016;**68**:1743–50. https://doi.org/10.1002/acr.22897.
143. Paschos NK, Lim N, Hu JC, Athanasiou KA. Functional properties of native and tissue-engineered cartilage toward understanding the pathogenesis of chondral lesions at the knee: A bovine cadaveric study. *Journal of Orthopaedic Research* 2017;**35**:2452–64. https://doi.org/10.1002/jor.23558.
144. Bielajew BJ, Hu JC, Athanasiou KA. Methodology to Quantify Collagen Subtypes and Crosslinks: Application in Minipig Cartilages. *Cartilage* 2021;**13**:1742S-1754S. https://doi.org/10.1177/19476035211060508.
145. Kempson GE, Muir H, Pollard C, Tuke M. The tensile properties of the cartilage of human femoral condyles related to the content of collagen and glycosaminoglycans. *Biochimica et Biophysica Acta (BBA) - General Subjects* 1973;**297**:456–72. https://doi.org/10.1016/0304-4165(73)90093-7.
146. Athanasiou KA, Rosenwasser MP, Buckwalter JA, Malinin TI, Mow VC. Interspecies comparisons of in situ intrinsic mechanical properties of distal femoral cartilage. *Journal of Orthopaedic Research* 1991;**9**:330–40. https://doi.org/10.1002/jor.1100090304.
147. Armstrong CG, Mow VC. Variations in the intrinsic mechanical properties of human articular cartilage with age, degeneration, and water content. *JBJS* 1982;**64**:.
148. Sweigart MA, Zhu CF, Burt DM, DeHoll PD, Agrawal CM, Clanton TO, *et al.* Intraspecies and interspecies comparison of the compressive properties of the medial meniscus. *Ann Biomed Eng* 2004;**32**:1569–79.
149. Sweigart MA, Athanasiou KA. Biomechanical characteristics of the normal medial and lateral porcine knee menisci. *Proc Inst Mech Eng [H]* 2005;**219**:53–62.
150. Athanasiou KA, Agarwal A, Dzida FJ. Comparative study of the intrinsic mechanical properties of the human acetabular and femoral head cartilage. *J Orthop Res* 1994;**12**:340–9. https://doi.org/10.1002/jor.1100120306.

151. Oliveria SA, Felson DT, Reed JI, Cirillo PA, Walker AM. Incidence of symptomatic hand, hip, and knee osteoarthritis among patients in a health maintenance organization. *Arthritis Rheum* 1995;**38**:1134–41. https://doi.org/10.1002/art.1780380817.

152. Ferguson RJ, Palmer AJR, Taylor A, Porter ML, Malchau H, Glyn-Jones S. Hip replacement. *The Lancet* 2018;**392**:1662–71. https://doi.org/10.1016/S0140-6736(18)31777-X.

153. Fackler NP, Donahue RP, Bielajew BJ, Amirhekmat A, Hu JC, Athanasiou KA, *et al.* Characterization of the Age-Related Differences in Porcine Acetabulum and Femoral Head Articular Cartilage. *Cartilage* 2023:19476035231214724. https://doi.org/10.1177/19476035231214724.

154. Ferguson SJ, Bryant JT, Ito K. The material properties of the bovine acetabular labrum. *Journal of Orthopaedic Research* 2001;**19**:887–96. https://doi.org/10.1016/S0736-0266(01)00007-9.

155. Ishiko T, Naito M, Moriyama S. Tensile properties of the human acetabular labrum—the first report. *Journal of Orthopaedic Research* 2005;**23**:1448–53. https://doi.org/10.1016/j.orthres.2004.08.025.1100230630.

156. Magnussen RA, Guilak F, Vail TP. Cartilage degeneration in post-collapse cases of osteonecrosis of the human femoral head: Altered mechanical properties in tension, compression, and shear. *Journal of Orthopaedic Research* 2005;**23**:576–83. https://doi.org/10.1016/j.orthres.2004.12.006.

157. Herrera-Pérez M, Valderrabano V, Godoy-Santos AL, de César Netto C, González-Martín D, Tejero S. Ankle osteoarthritis: comprehensive review and treatment algorithm proposal. *EFORT Open Rev* 2022;**7**:448–59. https://doi.org/10.1530/EOR-21-0117.

158. Athanasiou KA, Niederauer GG, Schenck RC, Schenck Jr. RC. Biomechanical topography of human ankle cartilage. *Ann Biomed Eng* 1995;**23**:697–704. https://doi.org/10.1007/BF02584467.

159. Paschos NK, Makris EA, Hu JC, Athanasiou KA. Topographic variations in biomechanical and biochemical properties in the ankle joint: An in vitro bovine study evaluating native and engineered cartilage. *Arthroscopy - Journal of Arthroscopic and Related Surgery* 2014;**30**:1317–26. https://doi.org/10.1016/j.arthro.2014.05.025.

160. Valesan LF, Da-Cas CD, Réus JC, Denardin ACS, Garanhani RR, Bonotto D, *et al.* Prevalence of temporomandibular joint disorders: a systematic review and meta-analysis. *Clin Oral Investig* 2021;**25**:441–53. https://doi.org/10.1007/s00784-020-03710-w.

161. Detamore MS, Athanasiou KA. Motivation, Characterization, and Strategy for Tissue Engineering the Temporomandibular Joint Disc. *Tissue Eng* 2003;**9**:1065–87. https://doi.org/10.1089/10763270360727991.

162. Kalpakci KN, Willard VP, Wong ME, Athanasiou KA. An interspecies comparison of the temporomandibular joint disc. *J Dent Res* 2011;**90**:193–8. https://doi.org/10.1177/0022034510381501.

163. Vapniarsky N, Aryaei A, Arzi B, Hatcher DC, Hu JC, Athanasiou KA. The Yucatan Minipig Temporomandibular Joint Disc Structure–Function Relationships Support Its Suitability for Human Comparative Studies. *Tissue Eng Part C Methods* 2017;**23**:700–9. https://doi.org/10.1089/ten.tec.2017.0149.

164. Singh M, Detamore MS. Biomechanical properties of the mandibular condylar cartilage and their relevance to the TMJ disc. *J Biomech* 2009;**42**:405–17. https://doi.org/10.1016/j.jbiomech.2008.12.012.

165. Kuroda S, Tanimoto K, Izawa T, Fujihara S, Koolstra JH, Tanaka E. Biomechanical and biochemical characteristics of the mandibular condylar cartilage. *Osteoarthritis Cartilage* 2009;**17**:1408–15. https://doi.org/10.1016/j.joca.2009.04.025.

166. Singh M, Detamore MS. Tensile properties of the mandibular condylar cartilage. *J Biomech Eng* 2008;**130**:. https://doi.org/10.1115/1.2838062.

167. Lu XL, Mow VC, Guo XE. Proteoglycans and mechanical behavior of condylar cartilage. *J Dent Res* 2009;**88**:244–8. https://doi.org/10.1177/0022034508330432.
168. Allen KD, Athanasiou KA. Viscoelastic characterization of the porcine temporomandibular joint disc under unconfined compression. *J Biomech* 2006;**39**:312–22. https://doi.org/10.1016/j.jbiomech.2004.11.012.
169. Teraguchi M, Yoshimura N, Hashizume H, Muraki S, Yamada H, Minamide A, *et al.* Prevalence and distribution of intervertebral disc degeneration over the entire spine in a population-based cohort: The Wakayama Spine Study. *Osteoarthritis Cartilage* 2014;**22**:104–10. https://doi.org/10.1016/j.joca.2013.10.019.
170. Newell N, Little JP, Christou A, Adams MA, Adam CJ, Masouros SD. Biomechanics of the human intervertebral disc: A review of testing techniques and results. *J Mech Behav Biomed Mater* 2017:420–34. https://doi.org/10.1016/j.jmbbm.2017.01.037.
171. Elliott DM, Yerramalli CS, Auerbach JD. *Biomechanics of the Intervertebral Disc INTRODUCTION: STRUCTURE AND FUNCTION OF DISC*. n.d.
172. Johannessen W, Elliott DM. *Effects of Degeneration on the Biphasic Material Properties of Human Nucleus Pulposus in Confined Compression*. vol. 30. n.d.
173. Newell N, Little JP, Christou A, Adams MA, Adam CJ, Masouros SD. Biomechanics of the human intervertebral disc: A review of testing techniques and results. *J Mech Behav Biomed Mater* 2017:420–34. https://doi.org/10.1016/j.jmbbm.2017.01.037.
174. O'Leary SA, Paschos NK, Link JM, Klineberg EO, Hu JC, Athanasiou KA. Facet Joints of the Spine: Structure–Function Relationships, Problems and Treatments, and the Potential for Regeneration. *Annu Rev Biomed Eng* 2018;**20**:145–70. https://doi.org/10.1146/annurev-bioeng-062117-120924.
175. O'Leary SA, Link JM, Klineberg EO, Hu JC, Athanasiou KA. Characterization of facet joint cartilage properties in the human and interspecies comparisons. *Acta Biomater* 2017;**54**:367–76. https://doi.org/10.1016/j.actbio.2017.03.017.
176. Eubanks JD, Lee MJ, Cassinelli E, Ahn NU. Prevalence of lumbar facet arthrosis and its relationship to age, sex, and race: an anatomic study of cadaveric specimens. *Spine (Phila Pa 1976)* 2007;**32**:2058–62. https://doi.org/10.1097/BRS.0b013e318145a3a9.
177. Glaser J, Kreiner S. Facet Joint Interventions. *North American Spine Society* 2016.
178. Gellhorn AC, Katz JN, Suri P. Osteoarthritis of the spine: the facet joints. *Nat Rev Rheumatol* 2013;**9**:216–24. https://doi.org/10.1038/nrrheum.2012.199.
179. O'Leary SA, White JL, Hu JC, Athanasiou KA. Biochemical and biomechanical characterisation of equine cervical facet joint cartilage. *Equine Vet J* 2018. https://doi.org/10.1111/evj.12845.
180. Nordberg RC, Espinosa MG, Hu JC, Athanasiou KA. A Tribological Comparison of Facet Joint, Sacroiliac Joint, and Knee Cartilage in the Yucatan Minipig. *Cartilage* 2021:19476035211021904. https://doi.org/10.1177/19476035211021906.
181. Nordberg RC, Kim AN, Hight JM, Meka RS, Elder BD, Hu JC, *et al.* Biochemical and biomechanical characterization of the cervical, thoracic, and lumbar facet joint cartilage in the Yucatan minipig. *J Biomech* 2022;**142**:111238. https://doi.org/10.1016/J.JBIOMECH.2022.111238.
182. Gupta S, Xiao R, Fainor M, Mauck RL, Smith HE, Gullbrand SE. Level dependent alterations in human facet cartilage mechanics and bone morphometry with spine degeneration. *Journal of Orthopaedic Research* 2023;**41**:674–83. https://doi.org/10.1002/jor.25407.
183. Sembrano JN, Polly DW. *How Often Is Low Back Pain Not Coming From the Back?*. vol. 34. n.d.
184. BERNARD TNJR, KIRKALDY-WILLIS WH. Recognizing Specific Characteristics of Nonspecific Low Back Pain. *Clin Orthop Relat Res* 1987;**217**:.

185. Kiapour A, Joukar A, Elgafy H, Erbulut DU, Agarwal AK, Goel VK. Biomechanics of the sacroiliac joint: Anatomy, function, biomechanics, sexual dimorphism, and causes of pain. *Int J Spine Surg* 2020;**14**:S3–13. https://doi.org/10.14444/6077.
186. Nordberg RC, Hight JM, Kim AN, Meka RS, Elder BD, Hu JC, *et al.* Biomechanical, biochemical, and histological characterization of sacroiliac joint cartilage in the Yucatan minipig. *J Mech Behav Biomed Mater* 2024;**157**:106658. https://doi.org/10.1016/j.jmbbm.2024.106658.
187. Athanasiou KA, Liu GT, Lavery LA, Lanctot DR, Schenck Jr. RC. Biomechanical topography of human articular cartilage in the first metatarsophalangeal joint. *Clin Orthop Relat Res* 1998:269–81.
188. Liu GT, Lavery LA, Schenck RC, Lanctot DR, Zhu CF, Athanasiou KA. Human articular cartilage biomechanics of the second metatarsal intermediate cuneiform joint. *The Journal of Foot and Ankle Surgery* 1997;**36**:367–74. https://doi.org/10.1016/S1067-2516(97)80039-7.
189. Korvick D, Athanasiou K. Variations in the mechanical properties of cartilage from the canine scapulohumeral joint. *Am J Vet Res* 1997;**58**:949–53. https://doi.org/10.2460/ajvr.1997.58.09.949.
190. Schenck Jr. RC, Athanasiou KA, Constantinides G, Gomez E. A biomechanical analysis of articular cartilage of the human elbow and a potential relationship to osteochondritis dissecans. *Clin Orthop Relat Res* 1994:305–12.
191. Luquetti D V., Heike CL, Hing A V., Cunningham ML, Cox TC. Microtia: Epidemiology and genetics. *Am J Med Genet A* 2012:124–39. https://doi.org/10.1002/ajmg.a.34352.
192. Khan N, Zaki DP, Brown WE, Halaseh FF, Willette D, Ziegler M, *et al.* Tissue Engineering Auricular Cartilage: A Review of Auricular Cartilage Characteristics and Current Techniques for Auricular Reconstruction. *Journal of Craniofacial Surgery* 2024;**35**:768–78. https://doi.org/10.1097/SCS.0000000000010015.
193. Nimeskern L, Pleumeekers MM, Pawson DJ, Koevoet WLM, Lehtoviita I, Soyka MB, *et al.* Mechanical and biochemical mapping of human auricular cartilage for reliable assessment of tissue-engineered constructs. *J Biomech* 2015;**48**:1721–9. https://doi.org/10.1016/j.jbiomech.2015.05.019.
194. Bos EJ, Pluemeekers M, Helder M, Kuzmin N, van der Laan K, Groot ML, *et al.* Structural and mechanical comparison of human ear, alar, and septal cartilage. *Plast Reconstr Surg Glob Open* 2018;**6**:. https://doi.org/10.1097/GOX.0000000000001610.
195. Cohen BP, Hooper RC, Puetzer JL, Nordberg R, Asanbe O, Hernandez KA, *et al.* Long-Term Morphological and Microarchitectural Stability of Tissue-Engineered, Patient-Specific Auricles In Vivo. *Tissue EngineeringPart A* 2016;**22**:461–8. https://doi.org/10.1089/ten.TEA.2015.0323 [doi].
196. Griffin MF, Premakumar Y, Seifalian AM, Szarko M, Butler PEM. Biomechanical Characterisation of the Human Auricular Cartilages; Implications for Tissue Engineering. *Ann Biomed Eng* 2016;**44**:3460–7. https://doi.org/10.1007/s10439-016-1688-1.
197. Park SS, Jin HR, Chi DH, Taylor RS. Characteristics of tissue-engineered cartilage from human auricular chondrocytes. *Biomaterials* 2004;**25**:2363–9.
198. Nayyer L, Birchall M, Seifalian AM, Jell G. Design and development of nanocomposite scaffolds for auricular reconstruction. *Nanomedicine* 2014;**10**:235–46. https://doi.org/10.1016/j.nano.2013.06.006.
199. Lau AG, Kindig MW, Kent RW. Morphology, distribution, mineral density and volume fraction of human calcified costal cartilage. *Acta Biomater* 2011;**7**:1202–9. https://doi.org/10.1016/j.actbio.2010.10.019.
200. Palacios JF, Hazkour N, Robinson E, Swami P, Smith L, Grande D, *et al.* Primary Ear Reconstruction Using Cadaveric Costal Cartilage. *Ann Plast Surg* 2023;**90**:S547–51. https://doi.org/10.1097/SAP.0000000000003518.

201. Kalmar CL, Carlson AR, Bartlett SP. Double-Reversed Costal Cartilage Graft for Nasal Reconstruction. *Plast Reconstr Surg* 2024;**154**:170e–4e. https://doi.org/10.1097/PRS.0000000000010776.
202. Nummela MT, Bensch F V., Pyhältö TT, Koskinen SK. Incidence and imaging findings of costal cartilage fractures in patients with blunt chest trauma: A retrospective review of 1461 consecutive whole-body ct examinations for trauma. *Radiology* 2018:696–704. https://doi.org/10.1148/radiol.2017162429.
203. Sinclair S, Walsh WR. Characterization of costal cartilage allografts. *ANZ J Surg* 2022;**92**:2274–9. https://doi.org/10.1111/ans.17967.
204. Tocchioni F, Ghionzoli M, Calosi L, Guasti D, Romagnoli P, Messineo A. Rib cartilage characterization in patients affected by pectus excavatum. *Anatomical Record* 2013;**296**:1813–20. https://doi.org/10.1002/ar.22824.
205. Boazak EM, Auguste DT. Trachea Mechanics for Tissue Engineering Design. *ACS Biomater Sci Eng* 2018:1272–84. https://doi.org/10.1021/acsbiomaterials.7b00738.
206. Hoffman B, Martin M, Brown BN, Bonassar LJ, Cheetham J. Biomechanical and biochemical characterization of porcine tracheal cartilage. *Laryngoscope* 2016;**126**:E325–31. https://doi.org/10.1002/lary.25861.
207. Safshekan F, Tafazzoli-Shadpour M, Abdouss M, Shadmehr MB. Mechanical characterization and constitutive modeling of human trachea: Age and gender dependency. *Materials* 2016;**9**:. https://doi.org/10.3390/ma9060456.
208. Wang X, Li Z, Wang C, Bai H, Wang Z, Liu Y, *et al.* Enlightenment of Growth Plate Regeneration Based on Cartilage Repair Theory: A Review. *Front Bioeng Biotechnol* 2021;**9**:462. https://doi.org/10.3389/FBIOE.2021.654087/BIBTEX.
209. Dover C, Kiely N. Growth plate injuries and management. *Orthop Trauma* 2015;**29**:261–7. https://doi.org/10.1016/j.mporth.2015.04.002.
210. Cohen B, Chorney GS, Phillips DP, Dick HM, Buckwalter JA, Ratcliffe A, *et al.* The microstructural tensile properties and biochemical composition of the bovine distal femoral growth plate. *Journal of Orthopaedic Research* 1992;**10**:263–75. https://doi.org/10.1002/jor.1100100214.
211. Cohen B, Chorney GS, Phillips DP, Dick HM, Mow VC. Compressive stress-relaxation behavior of bovine growth plate may be described by the nonlinear biphasic theory. *Journal of Orthopaedic Research* 1994;**12**:804–13. https://doi.org/10.1002/jor.1100120608.
212. Sergerie K, Lacoursière MO, Lévesque M, Villemure I. Mechanical properties of the porcine growth plate and its three zones from unconfined compression tests. *J Biomech* 2009;**42**:510–6. https://doi.org/10.1016/j.jbiomech.2008.11.026.
213. Eckstein KN, Thomas SM, Scott AK, Neu CP, Hadley-Miller NA, Payne KA, *et al.* The heterogeneous mechanical properties of adolescent growth plate cartilage: A study in rabbit. *J Mech Behav Biomed Mater* 2022;**128**:. https://doi.org/10.1016/j.jmbbm.2022.105102.
214. Athanasiou KA, Eswaramoorthy R, Hadidi P, Hu JC. Self-organization and the self-assembling process in tissue engineering. *Annu Rev Biomed Eng* 2013;**15**:115–36. https://doi.org/10.1146/annurev-bioeng-071812-152423.
215. Tsujii A, Ohori T, Hanai H, Nakamura N. Next generation approaches for cartilage repair and joint preservation. *Journal of Cartilage and Joint Preservation* 2024. https://doi.org/10.1016/j.jcjp.2024.100177.
216. Obana KK, Schallmo MS, Hong IS, Ahmad CS, Moorman CT, Trofa DP, *et al.* Current Trends in Orthobiologics: An 11-Year Review of the Orthopaedic Literature. *Am J Sports Med* 2022;**50**:3121–9. https://doi.org/10.1177/03635465211037343.
217. Rodeo SA. Orthobiologics: Current Status in 2023 and Future Outlook. *J Am Acad Orthop Surg* 2023;**31**:604–13. https://doi.org/10.5435/JAAOS-D-22-00808.

218. Shapiro SA, Kazmerchak SE, Heckman MG, Zubair AC, O'Connor MI. A Prospective, Single-Blind, Placebo-Controlled Trial of Bone Marrow Aspirate Concentrate for Knee Osteoarthritis. *Am J Sports Med* 2017;**45**:82–90. https://doi.org/10.1177/0363546516662455.
219. Phinney DG, Pittenger MF. Concise Review: MSC-Derived Exosomes for Cell-Free Therapy. *Stem Cells* 2017;**35**:851–8. https://doi.org/10.1002/stem.2575.
220. Fice MP, Miller JC, Christian R, Hannon CP, Smyth N, Murawski CD, *et al.* The Role of Platelet-Rich Plasma in Cartilage Pathology: An Updated Systematic Review of the Basic Science Evidence. *Arthroscopy - Journal of Arthroscopic and Related Surgery* 2019:961–976.e3. https://doi.org/10.1016/j.arthro.2018.10.125.
221. Kim SH, Ha C-W, Park Y-B, Nam E, Lee J-E, Lee H-J. Intra-articular injection of mesenchymal stem cells for clinical outcomes and cartilage repair in osteoarthritis of the knee: a meta-analysis of randomized controlled trials. *Arch Orthop Trauma Surg* 2019;**139**:971–80. https://doi.org/10.1007/s00402-019-03140-8.
222. Previtali D, Merli G, Di Laura Frattura G, Candrian C, Zaffagnini S, Filardo G. The Long-Lasting Effects of 'Placebo Injections' in Knee Osteoarthritis: A Meta-Analysis. *Cartilage* 2021;**13**:185S-196S. https://doi.org/10.1177/1947603520906597.
223. Dai W, Leng X, Wang J, Shi Z, Cheng J, Hu X, *et al.* Intra-Articular Mesenchymal Stromal Cell Injections Are No Different From Placebo in the Treatment of Knee Osteoarthritis: A Systematic Review and Meta-analysis of Randomized Controlled Trials. *Arthroscopy* 2021;**37**:340–58. https://doi.org/10.1016/j.arthro.2020.10.016.
224. Prodromidis AD, Charalambous CP, Moran E, Venkatesh R, Pandit H. The role of Platelet-Rich Plasma (PRP) intraarticular injections in restoring articular cartilage of osteoarthritic knees. A systematic review and meta-analysis. *Osteoarthr Cartil Open* 2022;**4**:100318. https://doi.org/10.1016/j.ocarto.2022.100318.
225. Lucidi GA, Grassi A, Agostinone P, Di Paolo S, Dal Fabbro G, D'Alberton C, *et al.* Risk Factors Affecting the Survival Rate of Collagen Meniscal Implant for Partial Meniscal Deficiency: An Analysis of 156 Consecutive Cases at a Mean 10 Years of Follow-up. *Am J Sports Med* 2022;**50**:2900–8. https://doi.org/10.1177/03635465221112635.
226. Pimenta L, Marchi L, Coutinho E, Oliveira L. Lessons Learned After 9 Years' Clinical Experience with 3 Different Nucleus Replacement Devices. *Semin Spine Surg* 2012;**24**:43–7. https://doi.org/10.1053/j.semss.2011.11.009.
227. Al Kadah B, Naumann A, Schneider M, Schick B, Linxweiler M, Papaspyrou G. Auricular reconstruction with polyethylene implants or silicone prosthesis: A single institution experience. *J Craniomaxillofac Surg* 2018;**46**:2150–6. https://doi.org/10.1016/j.jcms.2018.10.005.
228. Pot MW, van Kuppevelt TH, Gonzales VK, Buma P, IntHout J, de Vries RBM, *et al.* Augmented cartilage regeneration by implantation of cellular versus acellular implants after bone marrow stimulation: a systematic review and meta-analysis of animal studies. *PeerJ* 2017;**5**:e3927. https://doi.org/10.7717/peerj.3927.
229. Andriolo L, Reale D, Di Martino A, Boffa A, Zaffagnini S, Filardo G. Cell-Free Scaffolds in Cartilage Knee Surgery: A Systematic Review and Meta-Analysis of Clinical Evidence. *Cartilage* 2021;**12**:277–92. https://doi.org/10.1177/1947603519852406.
230. Ow ZGW, Cheang HLX, Koh JH, Koh JZE, Lim KK-L, Wang D, *et al.* Does the Choice of Acellular Scaffold and Augmentation With Bone Marrow Aspirate Concentrate Affect Short-term Outcomes in Cartilage Repair? A Systematic Review and Meta-analysis. *Am J Sports Med* 2023;**51**:1622–33. https://doi.org/10.1177/03635465211069565.
231. Nordberg RC, Bielajew BJ, Takahashi T, Dai S, Hu JC, Athanasiou KA. Recent advancements in cartilage tissue engineering innovation and translation. *Nat Rev Rheumatol* 2024;**20**:323–46. https://doi.org/10.1038/s41584-024-01118-4.

232. Nordberg RC, Otarola GA, Wang D, Hu JC, Athanasiou KA. Navigating regulatory pathways for translation of biologic cartilage repair products. *Sci Transl Med* 2022;**14**:eabp8163. https://doi.org/10.1126/SCITRANSLMED.ABP8163/SUPPL_FILE/SCITRANSLMED.ABP8163_SM.PDF.

233. U.S. Food and Drug Administration. Guidance for Industry: Preparation of IDEs and INDs for Products Intended to Repair or Replace Knee Cartilage. *Www.FDA.Gov* 2011:45.

234. Nordberg RC, Donahue RP, Espinosa MG, Salinas EY, Hu JC, Athanasiou KA. A cell bank paradigm for preclinical evaluation of an analogous cellular product for an allogeneic cell therapy. *Biofabrication* 2024;**16**:035024. https://doi.org/10.1088/1758-5090/ad4de2.

235. Vapniarsky N, Huwe LW, Arzi B, Houghton MK, Wong ME, Wilson JW, *et al.* Tissue engineering toward temporomandibular joint disc regeneration. *Sci Transl Med* 2018;**10**:eaaq1802. https://doi.org/10.1126/scitranslmed.aaq1802.

236. Kwon H, Brown WE, O'Leary SA, Hu JC, Athanasiou KA. Rejuvenation of extensively passaged human chondrocytes to engineer functional articular cartilage. *Biofabrication* 2021. https://doi.org/10.1088/1758-5090/abd9d9.

237. Takahashi T, Brown WE, Trinh CT, Hu JC, Athanasiou KA. Formation of functionally robust human neocartilage from multiple donors using highly expanded costochondral cells. *Biofabrication* 2025;**17**:. https://doi.org/10.1088/1758-5090/ade18c.

238. Takahashi T, Donahue RP, Nordberg RC, Hu JC, Currall SC, Athanasiou KA. Commercialization of regenerative-medicine therapies. *Nature Reviews Bioengineering* 2023;**1**:906–29. https://doi.org/10.1038/s44222-023-00095-9.

239. Lohan A, Marzahn U, El Sayed K, Bock C, Haisch A, Kohl B, *et al.* Heterotopic and orthotopic autologous chondrocyte implantation using a minipig chondral defect model. *Ann Anat* 2013;**195**:488–97. https://doi.org/10.1016/j.aanat.2013.04.009.

240. Kojima K, Bonassar LJ, Ignotz RA, Syed K, Cortiella J, Vacanti CA. Comparison of tracheal and nasal chondrocytes for tissue engineering of the trachea. *Ann Thorac Surg* 2003;**76**:1884–8. https://doi.org/10.1016/s0003-4975(03)01193-7.

241. Komura M, Komura H, Tanaka Y, Kanamori Y, Sugiyama M, Nakahara S, *et al.* Human tracheal chondrocytes as a cell source for augmenting stenotic tracheal segments: the first feasibility study in an in vivo culture system. *Pediatr Surg Int* 2008;**24**:1117–21. https://doi.org/10.1007/s00383-008-2218-5.

242. Brittberg M, Lindahl A, Nilsson A, Ohlsson C, Isaksson O, Peterson L. Treatment of deep cartilage defects in the knee with autologous chondrocyte transplantation. *N Engl J Med* 1994;**331**:889–95. https://doi.org/10.1056/NEJM199410063311401.

243. Sato M, Yamato M, Mitani G, Takagaki T, Hamahashi K, Nakamura Y, *et al.* Combined surgery and chondrocyte cell-sheet transplantation improves clinical and structural outcomes in knee osteoarthritis. *NPJ Regen Med* 2019;**4**:4. https://doi.org/10.1038/s41536-019-0069-4.

244. Adkisson HD, Martin JA, Amendola RL, Milliman C, Mauch KA, Katwal AB, *et al.* The Potential of Human Allogeneic Juvenile Chondrocytes for Restoration of Articular Cartilage. *Am J Sports Med* 2010;**38**:1324–33. https://doi.org/10.1177/0363546510361950.

245. Cavalli E, Levinson C, Hertl M, Broguiere N, Brück O, Mustjoki S, *et al.* Characterization of polydactyly chondrocytes and their use in cartilage engineering. *Sci Rep* 2019;**9**:. https://doi.org/10.1038/s41598-019-40575-w.

246. Maehara M, Sato M, Toyoda E, Takahashi T, Okada E, Kotoku T, *et al.* Characterization of polydactyly-derived chondrocyte sheets versus adult chondrocyte sheets for articular cartilage repair. *Inflamm Regen* 2017;**37**:. https://doi.org/10.1186/s41232-017-0053-6.

247. Kreuz PC, Gentili C, Samans B, Martinelli D, Krüger JP, Mittelmeier W, *et al.* Scaffold-assisted cartilage tissue engineering using infant chondrocytes from human hip cartilage. *Osteoarthritis Cartilage* 2013;**21**:1997–2005. https://doi.org/10.1016/j.joca.2013.09.007.

248. Hamahashi K, Toyoda E, Ishihara M, Mitani G, Takagaki T, Kaneshiro N, *et al.* Polydactyly-derived allogeneic chondrocyte cell-sheet transplantation with high tibial osteotomy as regenerative therapy for knee osteoarthritis. *NPJ Regen Med* 2022;**7**:71. https://doi.org/10.1038/s41536-022-00272-1.
249. Tay AG, Farhadi J, Suetterlin R, Pierer G, Heberer M, Martin I. Cell yield, proliferation, and postexpansion differentiation capacity of human ear, nasal, and rib chondrocytes. *Tissue Eng* 2004;**10**:762–70. https://doi.org/10.1089/1076327041348572.
250. Pelttari K, Pippenger B, Mumme M, Feliciano S, Scotti C, Mainil-Varlet P, *et al.* Adult human neural crest-derived cells for articular cartilage repair. *Sci Transl Med* 2014;**6**:251ra119. https://doi.org/10.1126/scitranslmed.3009688.
251. Rogina A, Pušić M, Štefan L, Ivković A, Urlić I, Ivanković M, *et al.* Characterization of Chitosan-Based Scaffolds Seeded with Sheep Nasal Chondrocytes for Cartilage Tissue Engineering. *Ann Biomed Eng* 2021;**49**:1572–86. https://doi.org/10.1007/s10439-020-02712-9.
252. Kojima K, Bonassar LJ, Roy AK, Vacanti CA, Cortiella J. Autologous tissue-engineered trachea with sheep nasal chondrocytes. *J Thorac Cardiovasc Surg* 2002;**123**:1177–84. https://doi.org/10.1067/mtc.2002.121161.
253. Fulco I, Miot S, Haug MD, Barbero A, Wixmerten A, Feliciano S, *et al.* Engineered autologous cartilage tissue for nasal reconstruction after tumour resection: an observational first-in-human trial. *Lancet* 2014;**384**:337–46. https://doi.org/10.1016/S0140-6736(14)60544-4.
254. Mumme M, Barbero A, Miot S, Wixmerten A, Feliciano S, Wolf F, *et al.* Nasal chondrocyte-based engineered autologous cartilage tissue for repair of articular cartilage defects: an observational first-in-human trial. *Lancet* 2016;**388**:1985–94. https://doi.org/10.1016/S0140-6736(16)31658-0.
255. Acevedo Rua L, Mumme M, Manferdini C, Darwiche S, Khalil A, Hilpert M, *et al.* Engineered nasal cartilage for the repair of osteoarthritic knee cartilage defects. *Sci Transl Med* 2021;**13**:eaaz4499. https://doi.org/10.1126/scitranslmed.aaz4499.
256. Mumme M, Wixmerten A, Ivkovic A, Peretti GM, Yilmaz T, Reppenhagen S, *et al.* Clinical relevance of engineered cartilage maturation in a randomized multicenter trial for articular cartilage repair. *Sci Transl Med* 2025;**17**:eads0848. https://doi.org/10.1126/scitranslmed.ads0848.
257. Anderson DEJ, Athanasiou KA. Passaged goat costal chondrocytes provide a feasible cell source for temporomandibular joint tissue engineering. *Ann Biomed Eng* 2008;**36**:1992–2001. https://doi.org/10.1007/s10439-008-9572-2.
258. Huwe LW, Sullan GK, Hu JC, Athanasiou KA. Using Costal Chondrocytes to Engineer Articular Cartilage with Applications of Passive Axial Compression and Bioactive Stimuli 2018;**24**:516–26. https://doi.org/10.1089/ten.tea.2017.0136.
259. Yoon K-H, Park J-Y, Lee J-Y, Lee E, Lee J, Kim S-G. Costal Chondrocyte–Derived Pellet-Type Autologous Chondrocyte Implantation for Treatment of Articular Cartilage Defect. *Am J Sports Med* 2020;**48**:1236–45. https://doi.org/10.1177/0363546520905565.
260. Yoon K-H, Lee J, Park J-Y. Costal Chondrocyte–Derived Pellet-Type Autologous Chondrocyte Implantation Versus Microfracture for the Treatment of Articular Cartilage Defects: A 5-Year Follow-up of a Prospective Randomized Trial. *Am J Sports Med* 2024;**52**:362–7. https://doi.org/10.1177/03635465231222797.
261. Donahue RP, Nordberg RC, Bielajew BJ, Hu JC, Athanasiou KA. The effect of neonatal, juvenile, and adult donors on rejuvenated neocartilage functional properties. *Tissue Eng Part A* 2021;**28**:383–93. https://doi.org/10.1089/ten.TEA.2021.0167.
262. Zheng K, Ma Y, Chiu C, Xue M, Zhang C, Du D. Enhanced articular cartilage regeneration using costal chondrocyte-derived scaffold-free tissue engineered constructs with ascorbic acid treatment. *J Orthop Translat* 2024;**45**:140–54. https://doi.org/10.1016/j.jot.2024.02.005.

263. Lavernia L, Brown WE, Wong BJF, Hu JC, Athanasiou KA. Toward tissue-engineering of nasal cartilages. *Acta Biomater* 2019:42–56. https://doi.org/10.1016/j.actbio.2019.02.025.
264. Kusuhara H, Isogai N, Enjo M, Otani H, Ikada Y, Jacquet R, *et al.* Tissue engineering a model for the human ear: assessment of size, shape, morphology, and gene expression following seeding of different chondrocytes. *Wound Repair Regen* 2009;**17**:136–46. https://doi.org/10.1111/j.1524-475X.2008.00451.x.
265. Walles T, Giere B, Macchiarini P, Mertsching H. Expansion of chondrocytes in a three-dimensional matrix for tracheal tissue engineering. *Ann Thorac Surg* 2004;**78**:444–8; discussion 448–9. https://doi.org/10.1016/j.athoracsur.2004.02.122.
266. Nomoto M, Nomoto Y, Tada Y, Tani A, Otsuki K, Suzuki R, *et al.* Bioengineered trachea using autologous chondrocytes for regeneration of tracheal cartilage in a rabbit model. *Laryngoscope* 2013;**123**:2195–201. https://doi.org/10.1002/lary.23784.
267. Pangborn CA, Athanasiou KA. Effects of growth factors on meniscal fibrochondrocytes. *Tissue Eng* n.d.;**11**:1141–8. https://doi.org/10.1089/ten.2005.11.1141.
268. Martinek V, Ueblacker P, Bräun K, Nitschke S, Mannhardt R, Specht K, *et al.* Second generation of meniscus transplantation: in-vivo study with tissue engineered meniscus replacement. *Arch Orthop Trauma Surg* 2006;**126**:228–34. https://doi.org/10.1007/s00402-005-0025-1.
269. Hoben GM, Hu JC, James RA, Athanasiou KA. Self-assembly of fibrochondrocytes and chondrocytes for tissue engineering of the knee meniscus. *Tissue Eng* 2007;**13**:939–46. https://doi.org/10.1089/ten.2006.0116.
270. Schoenfeld AJ, Jacquet R, Lowder E, Doherty A, Leeson MC, Landis WJ. Histochemical analyses of tissue-engineered human menisci. *Connect Tissue Res* 2009;**50**:307–14.
271. Gonzalez-Leon EA, Bielajew BJ, Hu JC, Athanasiou KA. Engineering self-assembled neomenisci through combination of matrix augmentation and directional remodeling. *Acta Biomater* 2020;**109**:73–81. https://doi.org/10.1016/j.actbio.2020.04.019.
272. Klarmann GJ, Gaston J, Ho VB. A review of strategies for development of tissue engineered meniscal implants. *Biomaterials and Biosystems* 2021;**4**:100026. https://doi.org/10.1016/j.bbiosy.2021.100026.
273. Bian Y, Cai X, Zhou R, Lv Z, Xu Y, Wang Y, *et al.* Advances in meniscus tissue engineering: Towards bridging the gaps from bench to bedside. *Biomaterials* 2025;**312**:122716. https://doi.org/10.1016/j.biomaterials.2024.122716.
274. Sato M, Asazuma T, Ishihara M, Ishihara M, Kikuchi T, Kikuchi M, *et al.* An experimental study of the regeneration of the intervertebral disc with an allograft of cultured annulus fibrosus cells using a tissue-engineering method. *Spine (Phila Pa 1976)* 2003;**28**:548–53. https://doi.org/10.1097/01.BRS.0000049909.09102.60.
275. Pan Y, Chu T, Dong S, Hao Y, Ren X, Wang J, *et al.* Cells scaffold complex for Intervertebral disc Anulus Fibrosus tissue engineering: in vitro culture and product analysis. *Mol Biol Rep* 2012;**39**:8581–94. https://doi.org/10.1007/s11033-012-1710-0.
276. Allen KD, Athanasiou KA. Effect of passage and topography on gene expression of temporomandibular joint disc cells. *Tissue Eng* 2007;**13**:101–10.
277. Lavi A, Pelled G, Tawackoli W, Casap N, Gazit D, Gazit Z. Isolation and characterization of mesenchymal stromal progenitors from the temporomandibular joint disc. *J Tissue Eng Regen Med* 2017;**11**:1553–61. https://doi.org/10.1002/term.2055.
278. Van Osch GJVM, Mandl EW, Jahr H, Koevoet W, Nolst-Trenité G, Verhaar JAN. Considerations on the use of ear chondrocytes as donor chondrocytes for cartilage tissue engineering. *Biorheology* 2004;**41**:411–21.
279. Bernstein JL, Cohen BP, Lin A, Harper A, Bonassar LJ, Spector JA. Tissue Engineering Auricular Cartilage Using Late Passage Human Auricular Chondrocytes. *Ann Plast Surg* 2018;**80**:S168–73. https://doi.org/10.1097/SAP.0000000000001400.

280. Yanaga H, Imai K, Fujimoto T, Yanaga K. Generating ears from cultured autologous auricular chondrocytes by using two-stage implantation in treatment of microtia. *Plast Reconstr Surg* 2009;**124**:817–25. https://doi.org/10.1097/PRS.0b013e3181b17c0e.
281. Yanaga H, Imai K, Koga M, Yanaga K. Cell-engineered human elastic chondrocytes regenerate natural scaffold in vitro and neocartilage with neoperichondrium in the human body post-transplantation. *Tissue Eng Part A* 2012;**18**:2020–9. https://doi.org/10.1089/ten.TEA.2011.0370.
282. Sakahara D, Yanaga H, Noto M, Fujimoto T, Imai K. Long-Term Clinical Results of Two-Stage Total Ear Reconstruction of Microtia Using Autologous Cell-Engineered Chondrocytes. *Plast Reconstr Surg* 2023;**151**:282e–7e. https://doi.org/10.1097/PRS.0000000000009854.
283. Zhou G, Jiang H, Yin Z, Liu Y, Zhang Q, Zhang C, *et al.* In Vitro Regeneration of Patient-specific Ear-shaped Cartilage and Its First Clinical Application for Auricular Reconstruction. *EBioMedicine* 2018;**28**:287–302. https://doi.org/10.1016/j.ebiom.2018.01.011.
284. Wu W, Chen F, Feng X, Liu Y, Mao T. Engineering cartilage tissues with the shape of human nasal alar by using chondrocyte macroaggregate--Experiment study in rabbit model. *J Biotechnol* 2007;**130**:75–84. https://doi.org/10.1016/j.jbiotec.2007.02.029.
285. Chang CS, Yang CY, Hsiao HY, Chen L, Chu IM, Cheng MH, *et al.* Cultivation of auricular chondrocytes in poly(ethylene glycol)/poly(ε-caprolactone) hydrogel for tracheal cartilage tissue engineering in a rabbit model. *Eur Cell Mater* 2018;**35**:350–64. https://doi.org/10.22203/eCM.v035a24.
286. Komura M, Komura H, Kanamori Y, Tanaka Y, Suzuki K, Sugiyama M, *et al.* An animal model study for tissue-engineered trachea fabricated from a biodegradable scaffold using chondrocytes to augment repair of tracheal stenosis. *J Pediatr Surg* 2008;**43**:2141–6. https://doi.org/10.1016/j.jpedsurg.2008.08.038.
287. Zha K, Li X, Yang Z, Tian G, Sun Z, Sui X, *et al.* Heterogeneity of mesenchymal stem cells in cartilage regeneration: from characterization to application. *NPJ Regen Med* 2021;**6**:14. https://doi.org/10.1038/s41536-021-00122-6.
288. Pittenger MF, Mackay AM, Beck SC, Jaiswal RK, Douglas R, Mosca JD, *et al.* Multilineage potential of adult human mesenchymal stem cells. *Science* 1999;**284**:143–7.
289. Herrmann M, Hildebrand M, Menzel U, Fahy N, Alini M, Lang S, *et al.* Phenotypic Characterization of Bone Marrow Mononuclear Cells and Derived Stromal Cell Populations from Human Iliac Crest, Vertebral Body and Femoral Head. *Int J Mol Sci* 2019;**20**:. https://doi.org/10.3390/ijms20143454.
290. Sanchez-Adams J, Athanasiou KA. Dermis isolated adult stem cells for cartilage tissue engineering. *Biomaterials* 2012;**33**:109–19. https://doi.org/10.1016/j.biomaterials.2011.09.038.
291. Vapniarsky N, Arzi B, Hu JC, Nolta JA, Athanasiou KA. Concise Review: Human Dermis as an Autologous Source of Stem Cells for Tissue Engineering and Regenerative Medicine. *Stem Cells Transl Med* 2015;**4**:1187–98. https://doi.org/10.5966/sctm.2015-0084.
292. Kwon H, Haudenschild AK, Brown WE, Vapniarsky N, Paschos NK, Arzi B, *et al.* Tissue engineering potential of human dermis-isolated adult stem cells from multiple anatomical locations. *PLoS One* 2017;**12**:e0182531. https://doi.org/10.1371/journal.pone.0182531.
293. Akgun I, Unlu MC, Erdal OA, Ogut T, Erturk M, Ovali E, *et al.* Matrix-induced autologous mesenchymal stem cell implantation versus matrix-induced autologous chondrocyte implantation in the treatment of chondral defects of the knee: a 2-year randomized study. *Arch Orthop Trauma Surg* 2015;**135**:251–63. https://doi.org/10.1007/s00402-014-2136-z.
294. Haleem AM, Singergy AA El, Sabry D, Atta HM, Rashed LA, Chu CR, *et al.* The Clinical Use of Human Culture-Expanded Autologous Bone Marrow Mesenchymal Stem Cells Transplanted on Platelet-Rich Fibrin Glue in the Treatment of Articular Cartilage Defects: A Pilot

Study and Preliminary Results. *Cartilage* 2010;**1**:253–61. https://doi.org/10.1177/194760351 0366027.

295. de Windt TS, Vonk LA, Slaper-Cortenbach ICM, van den Broek MPH, Nizak R, van Rijen MHP, *et al.* Allogeneic Mesenchymal Stem Cells Stimulate Cartilage Regeneration and Are Safe for Single-Stage Cartilage Repair in Humans upon Mixture with Recycled Autologous Chondrons. *Stem Cells* 2017;**35**:256–64. https://doi.org/10.1002/stem.2475.
296. de Windt TS, Vonk LA, Slaper-Cortenbach ICM, Nizak R, van Rijen MHP, Saris DBF. Allogeneic MSCs and Recycled Autologous Chondrons Mixed in a One-Stage Cartilage Cell Transplantion: A First-in-Man Trial in 35 Patients. *Stem Cells* 2017;**35**:1984–93. https://doi.org/10.1002/stem.2657.
297. Sadeghirad B, Rehman Y, Khosravirad A, Sofi-Mahmudi A, Zandieh S, Jomy J, *et al.* Mesenchymal stem cells for chronic knee pain secondary to osteoarthritis: A systematic review and meta-analysis of randomized trials. *Osteoarthritis Cartilage* 2024;**32**:1207–19. https://doi.org/10.1016/j.joca.2024.04.021.
298. Hohmann E, Keough N, Stokes D, Frank R, Rodeo S. Adipose- and bone marrow-derived stromal cells reduce pain in patients with knee osteoarthritis but do not substantially improve knee functionality: an updated systematic review and meta-analysis. *Eur J Orthop Surg Traumatol* 2025;**35**:214. https://doi.org/10.1007/s00590-025-04322-4.
299. Whitehouse MR, Howells NR, Parry MC, Austin E, Kafienah W, Brady K, *et al.* Repair of Torn Avascular Meniscal Cartilage Using Undifferentiated Autologous Mesenchymal Stem Cells: From In Vitro Optimization to a First-in-Human Study. *Stem Cells Transl Med* 2017;**6**:1237–48. https://doi.org/10.1002/sctm.16-0199.
300. Vangsness CT, Farr J, Boyd J, Dellaero DT, Mills CR, LeRoux-Williams M. Adult human mesenchymal stem cells delivered via intra-articular injection to the knee following partial medial meniscectomy A Randomized, Double-Blind, Controlled Study. *The Journal of Bone and Joint Surgery-American Volume* 2014;**96**:90–8. https://doi.org/10.2106/JBJS.M.00058.
301. Mei L, Shen B, Ling P, Liu S, Xue J, Liu F, *et al.* Culture-expanded allogenic adipose tissue-derived stem cells attenuate cartilage degeneration in an experimental rat osteoarthritis model. *PLoS One* 2017;**12**:e0176107. https://doi.org/10.1371/journal.pone.0176107.
302. Strioga M, Viswanathan S, Darinskas A, Slaby O, Michalek J. Same or not the same? Comparison of adipose tissue-derived versus bone marrow-derived mesenchymal stem and stromal cells. *Stem Cells Dev* 2012;**21**:2724–52. https://doi.org/10.1089/scd.2011.0722.
303. Zorzi AR, Amstalden EMI, Plepis AMG, Martins VCA, Ferretti M, Antonioli E, *et al.* Effect of human adipose tissue mesenchymal stem cells on the regeneration of ovine articular cartilage. *Int J Mol Sci* 2015;**16**:26813–31. https://doi.org/10.3390/ijms161125989.
304. Sun Q, Zhang L, Xu T, Ying J, Xia B, Jing H, *et al.* Combined use of adipose derived stem cells and TGF-β3 microspheres promotes articular cartilage regeneration in vivo. *Biotech Histochem* 2018;**93**:168–76. https://doi.org/10.1080/10520295.2017.1401663.
305. Kyriakidis T, Iosifidis M, Michalopoulos E, Melas I, Stavropoulos-Giokas C, Verdonk R. Good mid-term outcomes after adipose-derived culture-expanded mesenchymal stem cells implantation in knee focal cartilage defects. *Knee Surg Sports Traumatol Arthrosc* 2020;**28**:502–8. https://doi.org/10.1007/s00167-019-05688-9.
306. Tevlin R, desJardins-Park H, Huber J, DiIorio SE, Longaker MT, Wan DC. Musculoskeletal tissue engineering: Adipose derived stromal cell implementation for the treatment of osteoarthritis. *Biomaterials* 2022;**286**:121544. https://doi.org/10.1016/j.biomaterials.2022.121544.
307. Mautner K, Gottschalk M, Boden SD, Akard A, Bae WC, Black L, *et al.* Cell-based versus corticosteroid injections for knee pain in osteoarthritis: a randomized phase 3 trial. *Nat Med* 2023. https://doi.org/10.1038/s41591-023-02632-w.

308. Sakaguchi Y, Sekiya I, Yagishita K, Muneta T. Comparison of human stem cells derived from various mesenchymal tissues: superiority of synovium as a cell source. *Arthritis Rheum* 2005;**52**:2521–9. https://doi.org/10.1002/art.21212.
309. Ogata Y, Mabuchi Y, Yoshida M, Suto EG, Suzuki N, Muneta T, *et al.* Purified Human Synovium Mesenchymal Stem Cells as a Good Resource for Cartilage Regeneration. *PLoS One* 2015;**10**:e0129096. https://doi.org/10.1371/journal.pone.0129096.
310. Sekiya I, Koga H, Otabe K, Nakagawa Y, Katano H, Ozeki N, *et al.* Additional Use of Synovial Mesenchymal Stem Cell Transplantation Following Surgical Repair of a Complex Degenerative Tear of the Medial Meniscus of the Knee: A Case Report. *Cell Transplant* 2019;**28**:1445–54. https://doi.org/10.1177/0963689719863793.
311. Sekiya I, Koga H, Katano H, Mizuno M, Kohno Y, Otabe K, *et al.* Second-look arthroscopy after meniscus repair and synovial mesenchymal stem cell transplantation to treat degenerative flaps and radial tears of the medial meniscus: A case report. *J Orthop Sci* 2022;**27**:821–34. https://doi.org/10.1016/j.jos.2021.04.015.
312. Russo E, Caprnda M, Kruzliak P, Conaldi PG, Borlongan C V, La Rocca G. Umbilical Cord Mesenchymal Stromal Cells for Cartilage Regeneration Applications. *Stem Cells Int* 2022;**2022**:2454168. https://doi.org/10.1155/2022/2454168.
313. Ha C-W, Park Y-B, Chung J-Y, Park Y-G. Cartilage Repair Using Composites of Human Umbilical Cord Blood-Derived Mesenchymal Stem Cells and Hyaluronic Acid Hydrogel in a Minipig Model. *Stem Cells Transl Med* 2015;**4**:1044–51. https://doi.org/10.5966/sctm.2014-0264.
314. Park Y-B, Ha C-W, Lee C-H, Yoon YC, Park Y-G. Cartilage Regeneration in Osteoarthritic Patients by a Composite of Allogeneic Umbilical Cord Blood-Derived Mesenchymal Stem Cells and Hyaluronate Hydrogel: Results from a Clinical Trial for Safety and Proof-of-Concept with 7 Years of Extended Follow-Up. *Stem Cells Transl Med* 2017;**6**:613–21. https://doi.org/10.5966/sctm.2016-0157.
315. *MEDIPOST Completes Administration for Phase 3 Clinical Trial of CARTISTEM® for Knee Osteoarthritis in Japan.* https://En.Medi-Post.Co.Kr/Stem-Cell-Therapeutic/2024/11/37920 (2024). n.d.
316. Medipost Co Ltd. *NCT01733186: Evaluation of Safety and Exploratory Efficacy of CARTISTEM®, a Cell Therapy Product for Articular Cartilage Defects.* 2013.
317. Sung JJ. Medipost's 'Cartistem' ankle treatment additional indication rejected by Ministry of Food and Drug Safety, plans to re-challenge for Phase 3 clinical trial. https://www.ThebionewsNet/News/ArticleViewHtml?Idxno=11718 2024.
318. Murry CE, Keller G. Differentiation of Embryonic Stem Cells to Clinically Relevant Populations: Lessons from Embryonic Development. *Cell* 2008;**132**:661–80. https://doi.org/10.1016/j.cell.2008.02.008.
319. Oldershaw RA, Baxter MA, Lowe ET, Bates N, Grady LM, Soncin F, *et al.* Directed differentiation of human embryonic stem cells toward chondrocytes. *Nat Biotechnol* 2010;**28**:1187–94. https://doi.org/10.1038/nbt.1683.
320. Yamashita A, Morioka M, Yahara Y, Okada M, Kobayashi T, Kuriyama S, *et al.* Generation of scaffoldless hyaline cartilaginous tissue from human iPSCs. *Stem Cell Reports* 2015;**4**:404–18. https://doi.org/10.1016/j.stemcr.2015.01.016.
321. Craft AM, Ahmed N, Rockel JS, Baht GS, Alman BA, Kandel RA, *et al.* Specification of chondrocytes and cartilage tissues from embryonic stem cells. *Development* 2013;**140**:2597–610. https://doi.org/10.1242/dev.087890.
322. Craft AM, Rockel JS, Nartiss Y, Kandel RA, Alman BA, Keller GM. Generation of articular chondrocytes from human pluripotent stem cells. *Nat Biotechnol* 2015;**33**:638–45. https://doi.org/10.1038/nbt.3210.

323. Koay EJ, Hoben GMB, Athanasiou KA. Tissue engineering with chondrogenically differentiated human embryonic stem cells. *Stem Cells* 2007;**25**:2183–90. https://doi.org/10.1634/stemcells.2007-0105.
324. Koay EJ, Athanasiou KA. Development of serum-free, chemically defined conditions for human embryonic stem cell-derived fibrochondrogenesis. *Tissue Eng Part A* 2009;**15**:2249–57. https://doi.org/10.1089/ten.tea.2008.0320.
325. Barberi T, Willis LM, Socci ND, Studer L. Derivation of multipotent mesenchymal precursors from human embryonic stem cells. *PLoS Med* 2005;**2**:e161. https://doi.org/10.1371/journal.pmed.0020161.
326. Olivier EN, Rybicki AC, Bouhassira EE. Differentiation of human embryonic stem cells into bipotent mesenchymal stem cells. *Stem Cells* 2006;**24**:1914–22. https://doi.org/10.1634/stemcells.2005-0648.
327. Zhao J, Li S, Trilok S, Tanaka M, Jokubaitis-Jameson V, Wang B, *et al.* Small molecule-directed specification of sclerotome-like chondroprogenitors and induction of a somitic chondrogenesis program from embryonic stem cells. *Development* 2014;**141**:3848–58. https://doi.org/10.1242/dev.105981.
328. Takao T, Sato M, Fujisawa Y, Toyoda E, Yamada D, Hitsumoto Y, *et al.* A novel chondrocyte sheet fabrication using human-induced pluripotent stem cell-derived expandable limb-bud mesenchymal cells. *Stem Cell Res Ther* 2023;**14**:34. https://doi.org/10.1186/s13287-023-03252-4.
329. Evans MJ, Kaufman MH. Establishment in culture of pluripotential cells from mouse embryos. *Nature* 1981;**292**:154–6. https://doi.org/10.1038/292154a0.
330. Thomson JA, Itskovitz-Eldor J, Shapiro SS, Waknitz MA, Swiergiel JJ, Marshall VS, *et al.* Embryonic stem cell lines derived from human blastocysts. *Science* 1998;**282**:1145–7. https://doi.org/10.1126/science.282.5391.1145.
331. Hwang NS, Varghese S, Lee HJ, Zhang Z, Ye Z, Bae J, *et al.* In vivo commitment and functional tissue regeneration using human embryonic stem cell-derived mesenchymal cells. *Proc Natl Acad Sci U S A* 2008;**105**:20641–6. https://doi.org/10.1073/pnas.0809680106.
332. Toh WS, Lee EH, Guo X-M, Chan JKY, Yeow CH, Choo AB, *et al.* Cartilage repair using hyaluronan hydrogel-encapsulated human embryonic stem cell-derived chondrogenic cells. *Biomaterials* 2010;**31**:6968–80. https://doi.org/10.1016/j.biomaterials.2010.05.064.
333. Hoben GM, Willard VP, Athanasiou KA. Fibrochondrogenesis of hESCs: growth factor combinations and cocultures. *Stem Cells Dev* 2009;**18**:283–92. https://doi.org/10.1089/scd.2008.0024.
334. Griffith LA, Arnold KM, Sengers BG, Tare RS, Houghton FD. A scaffold-free approach to cartilage tissue generation using human embryonic stem cells. *Sci Rep* 2021;**11**:. https://doi.org/10.1038/s41598-021-97934-9.
335. Petrigliano FA, Liu NQ, Lee S, Tassey J, Sarkar A, Lin Y, *et al.* Long-term repair of porcine articular cartilage using cryopreservable, clinically compatible human embryonic stem cell-derived chondrocytes. *NPJ Regen Med* 2021;**6**:77. https://doi.org/10.1038/s41536-021-00187-3.
336. Takahashi K, Yamanaka S. Induction of pluripotent stem cells from mouse embryonic and adult fibroblast cultures by defined factors. *Cell* 2006;**126**:663–76. https://doi.org/10.1016/j.cell.2006.07.024.
337. Takahashi K, Tanabe K, Ohnuki M, Narita M, Ichisaka T, Tomoda K, *et al.* Induction of pluripotent stem cells from adult human fibroblasts by defined factors. *Cell* 2007;**131**:861–72. https://doi.org/10.1016/j.cell.2007.11.019.

338. Umeda K, Zhao J, Simmons P, Stanley E, Elefanty A, Nakayama N. Human chondrogenic paraxial mesoderm, directed specification and prospective isolation from pluripotent stem cells. *Sci Rep* 2012;**2**:455. https://doi.org/10.1038/srep00455.
339. Fukuta M, Nakai Y, Kirino K, Nakagawa M, Sekiguchi K, Nagata S, *et al.* Derivation of mesenchymal stromal cells from pluripotent stem cells through a neural crest lineage using small molecule compounds with defined media. *PLoS One* 2014;**9**:e112291. https://doi.org/10.1371/journal.pone.0112291.
340. Abe K, Yamashita A, Morioka M, Horike N, Takei Y, Koyamatsu S, *et al.* Engraftment of allogeneic iPS cell-derived cartilage organoid in a primate model of articular cartilage defect. *Nat Commun* 2023;**14**:804. https://doi.org/10.1038/s41467-023-36408-0.
341. Clinical Research Submission and Publication System. https://Jrct.Mhlw.Go.Jp/En-Latest-Detail/JRCTa050190104. n.d.
342. Koehler KR, Nie J, Longworth-Mills E, Liu X-P, Lee J, Holt JR, *et al.* Generation of inner ear organoids containing functional hair cells from human pluripotent stem cells. *Nat Biotechnol* 2017;**35**:583–9. https://doi.org/10.1038/nbt.3840.
343. Imaizumi M, Nomoto Y, Sato Y, Sugino T, Miyake M, Wada I, *et al.* Evaluation of the use of induced pluripotent stem cells (iPSCs) for the regeneration of tracheal cartilage. *Cell Transplant* 2013;**22**:341–53. https://doi.org/10.3727/096368912X653147.
344. Kim IG, Park SA, Lee S-H, Choi JS, Cho H, Lee SJ, *et al.* Transplantation of a 3D-printed tracheal graft combined with iPS cell-derived MSCs and chondrocytes. *Sci Rep* 2020;**10**:4326. https://doi.org/10.1038/s41598-020-61405-4.
345. Ali EAM, Smaida R, Meyer M, Ou W, Li Z, Han Z, *et al.* iPSCs chondrogenic differentiation for personalized regenerative medicine: a literature review. *Stem Cell Res Ther* 2024;**15**:185. https://doi.org/10.1186/s13287-024-03794-1.
346. Yamashita A, Tamamura Y, Morioka M, Karagiannis P, Shima N, Tsumaki N. Considerations in hiPSC-derived cartilage for articular cartilage repair. *Inflamm Regen* 2018;**38**:17. https://doi.org/10.1186/s41232-018-0075-8.
347. Hanatani T, Takasu N. CiRA iPSC seed stocks (CiRA's iPSC Stock Project). *Stem Cell Res* 2020;**50**:102033. https://doi.org/10.1016/j.scr.2020.102033.
348. Kyoto University. *iPS Cells: The Past Ten Years and the Future.* http://Www.Cira.Kyoto-u.Ac.Jp/e. n.d.
349. Vinod E, Padmaja K, Ramasamy B, Sathishkumar S. Systematic review of articular cartilage derived chondroprogenitors for cartilage repair in animal models. *J Orthop* 2023;**35**:43–53. https://doi.org/10.1016/j.jor.2022.10.012.
350. Dowthwaite GP, Bishop JC, Redman SN, Khan IM, Rooney P, Evans DJR, *et al.* The surface of articular cartilage contains a progenitor cell population. *J Cell Sci* 2004;**117**:889–97. https://doi.org/10.1242/jcs.00912.
351. Williams R, Khan IM, Richardson K, Nelson L, McCarthy HE, Analbelsi T, *et al.* Identification and Clonal Characterisation of a Progenitor Cell Sub-Population in Normal Human Articular Cartilage. *PLoS One* 2010;**5**:e13246. https://doi.org/10.1371/journal.pone.0013246.
352. Vinod E, Kachroo U, Rebekah G, Yadav BK, Ramasamy B. Characterization of human articular chondrocytes and chondroprogenitors derived from non-diseased and osteoarthritic knee joints to assess superiority for cell-based therapy. *Acta Histochem* 2020;**122**:151588. https://doi.org/10.1016/j.acthis.2020.151588.
353. Frisbie DD, McCarthy HE, Archer CW, Barrett MF, McIlwraith CW. Evaluation of articular cartilage progenitor cells for the repair of articular defects in an equine model. *Journal of Bone and Joint Surgery - American Volume* 2015;**97**:484–93. https://doi.org/10.2106/JBJS.N.00404.

354. Mancini IAD, Schmidt S, Brommer H, Pouran B, Schäfer S, Tessmar J, *et al.* A composite hydrogel-3D printed thermoplast osteochondral anchor as example for a zonal approach to cartilage repair: in vivo performance in a long-term equine model. *Biofabrication* 2020;**12**:35028. https://doi.org/10.1088/1758-5090/ab94ce.
355. Korpershoek J V, Rikkers M, de Windt TS, Tryfonidou MA, Saris DBF, Vonk LA. Selection of Highly Proliferative and Multipotent Meniscus Progenitors through Differential Adhesion to Fibronectin: A Novel Approach in Meniscus Tissue Engineering. *Int J Mol Sci* 2021;**22**:. https://doi.org/10.3390/ijms22168614.
356. Jayasuriya CT, Twomey-Kozak J, Newberry J, Desai S, Feltman P, Franco JR, *et al.* Human Cartilage-Derived Progenitors Resist Terminal Differentiation and Require CXCR4 Activation to Successfully Bridge Meniscus Tissue Tears. *Stem Cells* 2019;**37**:102–14. https://doi.org/10.1002/stem.2923.
357. Koelling S, Kruegel J, Irmer M, Path JR, Sadowski B, Miro X, *et al.* Migratory chondrogenic progenitor cells from repair tissue during the later stages of human osteoarthritis. *Cell Stem Cell* 2009;**4**:324–35. https://doi.org/10.1016/j.stem.2009.01.015.
358. Vinod E, Johnson NN, Kumar S, Amirtham SM, James JV, Livingston A, *et al.* Migratory chondroprogenitors retain superior intrinsic chondrogenic potential for regenerative cartilage repair as compared to human fibronectin derived chondroprogenitors. *Sci Rep* 2021;**11**:23685. https://doi.org/10.1038/s41598-021-03082-5.
359. Wang K, Li J, Li Z, Wang B, Qin Y, Zhang N, *et al.* Chondrogenic Progenitor Cells Exhibit Superiority Over Mesenchymal Stem Cells and Chondrocytes in Platelet-Rich Plasma Scaffold-Based Cartilage Regeneration. *Am J Sports Med* 2019;**47**:2200–15. https://doi.org/10.1177/0363546519854219.
360. Muhammad H, Schminke B, Bode C, Roth M, Albert J, von der Heyde S, *et al.* Human migratory meniscus progenitor cells are controlled via the TGF-β pathway. *Stem Cell Reports* 2014;**3**:789–803. https://doi.org/10.1016/j.stemcr.2014.08.010.
361. Carluccio S, Martinelli D, Palamà MEF, Pereira RC, Benelli R, Guijarro A, *et al.* Progenitor Cells Activated by Platelet Lysate in Human Articular Cartilage as a Tool for Future Cartilage Engineering and Reparative Strategies. *Cells* 2020;**9**:. https://doi.org/10.3390/cells9041052.
362. Vinod E, Parasuraman G, Lisha J J, Amirtham SM, Livingston A, Varghese JJ, *et al.* Human fetal cartilage-derived chondrocytes and chondroprogenitors display a greater commitment to chondrogenesis than adult cartilage resident cells. *PLoS One* 2023;**18**:e0285106. https://doi.org/10.1371/journal.pone.0285106.
363. Risbud M V, Guttapalli A, Tsai T-T, Lee JY, Danielson KG, Vaccaro AR, *et al.* Evidence for skeletal progenitor cells in the degenerate human intervertebral disc. *Spine (Phila Pa 1976)* 2007;**32**:2537–44. https://doi.org/10.1097/BRS.0b013e318158dea6.
364. Su X, Zuo W, Wu Z, Chen J, Wu N, Ma P, *et al.* CD146 as a new marker for an increased chondroprogenitor cell sub-population in the later stages of osteoarthritis. *J Orthop Res* 2015;**33**:84–91. https://doi.org/10.1002/jor.22731.
365. Hattori S, Oxford C, Reddi AH. Identification of superficial zone articular chondrocyte stem/progenitor cells. *Biochem Biophys Res Commun* 2007;**358**:99–103. https://doi.org/10.1016/j.bbrc.2007.04.142.
366. Murphy CL, Sambanis A. Effect of oxygen tension and alginate encapsulation on restoration of the differentiated phenotype of passaged chondrocytes. *Tissue Eng* 2001;**7**:791–803. https://doi.org/10.1089/107632701753337735.
367. Ma H-L, Hung S-C, Lin S-Y, Chen Y-L, Lo W-H. Chondrogenesis of human mesenchymal stem cells encapsulated in alginate beads. *J Biomed Mater Res A* 2003;**64**:273–81. https://doi.org/10.1002/jbm.a.10370.

368. Chu Y, Huang L, Hao W, Zhao T, Zhao H, Yang W, *et al.* Long-term stability, high strength, and 3D printable alginate hydrogel for cartilage tissue engineering application. *Biomed Mater* 2021;**16**:. https://doi.org/10.1088/1748-605X/ac2595.

369. Fragonas E, Valente M, Pozzi-Mucelli M, Toffanin R, Rizzo R, Silvestri F, *et al.* Articular cartilage repair in rabbits by using suspensions of allogenic chondrocytes in alginate. *Biomaterials* 2000;**21**:795–801. https://doi.org/10.1016/s0142-9612(99)00241-0.

370. Saris DB, Mukherjee N, Berglund LJ, Schultz FM, An KN, O'Driscoll SW. Dynamic pressure transmission through agarose gels. *Tissue Eng* 2000;**6**:531–7. https://doi.org/10.1089/107632700750022170.

371. Elder SH, Goldstein SA, Kimura JH, Soslowsky LJ, Spengler DM. Chondrocyte differentiation is modulated by frequency and duration of cyclic compressive loading. *Ann Biomed Eng* 2001;**29**:476–82. https://doi.org/10.1114/1.1376696.

372. Mauck RL, Soltz MA, Wang CC, Wong DD, Chao PH, Valhmu WB, *et al.* Functional tissue engineering of articular cartilage through dynamic loading of chondrocyte-seeded agarose gels. *J Biomech Eng* 2000;**122**:252–60. https://doi.org/10.1115/1.429656.

373. Selmi TAS, Verdonk P, Chambat P, Dubrana F, Potel J-F, Barnouin L, *et al.* Autologous chondrocyte implantation in a novel alginate-agarose hydrogel: outcome at two years. *J Bone Joint Surg Br* 2008;**90**:597–604. https://doi.org/10.1302/0301-620X.90B5.20360.

374. Clavé A, Potel J-F, Servien E, Neyret P, Dubrana F, Stindel E. Third-generation autologous chondrocyte implantation versus mosaicplasty for knee cartilage injury: 2-year randomized trial. *J Orthop Res* 2016;**34**:658–65. https://doi.org/10.1002/jor.23152.

375. Lynn AK, Yannas I V, Bonfield W. Antigenicity and immunogenicity of collagen. *J Biomed Mater Res B Appl Biomater* 2004;**71**:343–54. https://doi.org/10.1002/jbm.b.30096.

376. Freyria A-M, Ronzière M-C, Cortial D, Galois L, Hartmann D, Herbage D, *et al.* Comparative phenotypic analysis of articular chondrocytes cultured within type I or type II collagen scaffolds. *Tissue Eng Part A* 2009;**15**:1233–45. https://doi.org/10.1089/ten.tea.2008.0114.

377. Nürnberger S, Schneider C, van Osch GVM, Keibl C, Rieder B, Monforte X, *et al.* Repopulation of an auricular cartilage scaffold, AuriScaff, perforated with an enzyme combination. *Acta Biomater* 2019;**86**:207–22. https://doi.org/10.1016/j.actbio.2018.12.035.

378. Cohen BP, Bernstein JL, Morrison KA, Spector JA, Bonassar LJ. Tissue engineering the human auricle by auricular chondrocyte-mesenchymal stem cell co-implantation. *PLoS One* 2018;**13**:e0202356. https://doi.org/10.1371/journal.pone.0202356.

379. Hao T, Wen N, Cao J-K, Wang H-B, Lü S-H, Liu T, *et al.* The support of matrix accumulation and the promotion of sheep articular cartilage defects repair in vivo by chitosan hydrogels. *Osteoarthritis Cartilage* 2010;**18**:257–65. https://doi.org/10.1016/j.joca.2009.08.007.

380. Schmitt C, Radetzki F, Stirnweiss A, Mendel T, Ludtka C, Friedmann A, *et al.* Long-term pre-clinical evaluation of an injectable chitosan nanocellulose hydrogel with encapsulated adipose-derived stem cells in an ovine model for IVD regeneration. *J Tissue Eng Regen Med* 2021;**15**:660–73. https://doi.org/10.1002/term.3216.

381. Gunes OC, Kara A, Baysan G, Bugra Husemoglu R, Akokay P, Ziylan Albayrak A, *et al.* Fabrication of 3D Printed poly(lactic acid) strut and wet-electrospun cellulose nano fiber reinforced chitosan-collagen hydrogel composite scaffolds for meniscus tissue engineering. *J Biomater Appl* 2022;**37**:683–97. https://doi.org/10.1177/08853282221109339.

382. Sarem M, Moztarzadeh F, Mozafari M, Shastri VP. Optimization strategies on the structural modeling of gelatin/chitosan scaffolds to mimic human meniscus tissue. *Mater Sci Eng C Mater Biol Appl* 2013;**33**:4777–85. https://doi.org/10.1016/j.msec.2013.07.036.

383. Singh YP, Bandyopadhyay A, Dey S, Bhardwaj N, Mandal BB. Trends and advances in silk based 3D printing/bioprinting towards cartilage tissue engineering and regeneration. *Progress in Biomedical Engineering* 2024;**6**:022002. https://doi.org/10.1088/2516-1091/ad2d59.

384. Wang Y, Blasioli DJ, Kim H-J, Kim HS, Kaplan DL. Cartilage tissue engineering with silk scaffolds and human articular chondrocytes. *Biomaterials* 2006;**27**:4434–42. https://doi.org/10.1016/j.biomaterials.2006.03.050.
385. Kim SH, Yeon YK, Lee JM, Chao JR, Lee YJ, Seo YB, *et al.* Precisely printable and biocompatible silk fibroin bioink for digital light processing 3D printing. *Nat Commun* 2018;**9**:1620. https://doi.org/10.1038/s41467-018-03759-y.
386. Li Y, Chen M, Zhou W, Gao S, Luo X, Peng L, *et al.* Cell-free 3D wet-electrospun PCL/silk fibroin/Sr2+ scaffold promotes successful total meniscus regeneration in a rabbit model. *Acta Biomater* 2020;**113**:196–209. https://doi.org/10.1016/j.actbio.2020.06.017.
387. Bandyopadhyay A, Mandal BB. A three-dimensional printed silk-based biomimetic tri-layered meniscus for potential patient-specific implantation. *Biofabrication* 2019;**12**:015003. https://doi.org/10.1088/1758-5090/ab40fa.
388. Bhunia BK, Dey S, Bandyopadhyay A, Mandal BB. 3D printing of annulus fibrosus anatomical equivalents recapitulating angle-ply architecture for intervertebral disc replacement. *Appl Mater Today* 2021;**23**:101031. https://doi.org/10.1016/j.apmt.2021.101031.
389. Bhunia BK, Kaplan DL, Mandal BB. Silk-based multilayered angle-ply annulus fibrosus construct to recapitulate form and function of the intervertebral disc. *Proc Natl Acad Sci U S A* 2018;**115**:477–82. https://doi.org/10.1073/pnas.1715912115.
390. Singh YP, Bandyopadhyay A, Mandal BB. 3D Bioprinting Using Cross-Linker-Free Silk-Gelatin Bioink for Cartilage Tissue Engineering. *ACS Appl Mater Interfaces* 2019;**11**:33684–96. https://doi.org/10.1021/acsami.9b11644.
391. Sterodimas A, de Faria J. Human auricular tissue engineering in an immunocompetent animal model. *Aesthet Surg J* 2013;**33**:283–9. https://doi.org/10.1177/1090820X12472902.
392. Spotnitz WD. Fibrin Sealant: The Only Approved Hemostat, Sealant, and Adhesive-a Laboratory and Clinical Perspective. *ISRN Surg* 2014;**2014**:203943. https://doi.org/10.1155/2014/203943.
393. Kedage V V, Sanghavi SY, Badnre A, Desai NS. Autologous chondrocyte implantation (ACI): an innovative technique for articular cartilage defects. *J Clin Orthop Trauma* 2010;**1**:33–6. https://doi.org/10.1016/S0976-5662(11)60007-6.
394. Sanz-Horta R, Matesanz A, Gallardo A, Reinecke H, Jorcano JL, Acedo P, *et al.* Technological advances in fibrin for tissue engineering. *J Tissue Eng* 2023;**14**:20417314231190290. https://doi.org/10.1177/20417314231190288.
395. Kim MK, Choi SW, Kim SR, Oh IS, Won MH. Autologous chondrocyte implantation in the knee using fibrin. *Knee Surg Sports Traumatol Arthrosc* 2010;**18**:528–34. https://doi.org/10.1007/s00167-009-0905-y.
396. Choi N-Y, Kim B-W, Yeo W-J, Kim H-B, Suh D-S, Kim J-S, *et al.* Gel-type autologous chondrocyte (Chondron) implantation for treatment of articular cartilage defects of the knee. *BMC Musculoskelet Disord* 2010;**11**:103. https://doi.org/10.1186/1471-2474-11-103.
397. Tsanaktsidou E, Kammona O, Kiparissides C. Recent Developments in Hyaluronic Acid-Based Hydrogels for Cartilage Tissue Engineering Applications. *Polymers (Basel)* 2022;**14**:. https://doi.org/10.3390/polym14040839.
398. Dvořáková J, Kučera L, Kučera J, Švík K, Foglarová M, Muthný T, *et al.* Chondrogenic differentiation of mesenchymal stem cells in a hydrogel system based on an enzymatically crosslinked tyramine derivative of hyaluronan. *J Biomed Mater Res A* 2014;**102**:3523–30. https://doi.org/10.1002/jbm.a.35033.
399. Zhu D, Wang H, Trinh P, Heilshorn SC, Yang F. Elastin-like protein-hyaluronic acid (ELP-HA) hydrogels with decoupled mechanical and biochemical cues for cartilage regeneration. *Biomaterials* 2017;**127**:132–40. https://doi.org/10.1016/j.biomaterials.2017.02.010.

400. Chung C, Mesa J, Randolph MA, Yaremchuk M, Burdick JA. Influence of gel properties on neocartilage formation by auricular chondrocytes photoencapsulated in hyaluronic acid networks. *J Biomed Mater Res A* 2006;**77**:518–25. https://doi.org/10.1002/jbm.a.30660.
401. Eglin D, Mortisen D, Alini M. Degradation of synthetic polymeric scaffolds for bone and cartilage tissue repairs. *Soft Matter* 2009;**5**:938. https://doi.org/10.1039/b803718n.
402. Athanasiou KA, Darling EM, Hu JC. *Articular Cartilage Tissue Engineering*. Cham: Springer International Publishing; 2010.
403. Middleton JC, Tipton AJ. Synthetic biodegradable polymers as orthopedic devices. *Biomaterials* 2000;**21**:2335–46. https://doi.org/10.1016/s0142-9612(00)00101-0.
404. Athanasiou KA, Niederauer GG, Agrawal CM. Sterilization, toxicity, biocompatibility and clinical applications of polylactic acid/polyglycolic acid copolymers. *Biomaterials* 1996;**17**:93–102.
405. Athanasiou K, Korvick D, Schenck R. Biodegradable Implants for the Treatment of Osteochondral Defects in a Goat Model. https://HomeLiebertpubCom/Ten 1997;**3**:363–73. https://doi.org/10.1089/TEN.1997.3.363.
406. Daniels AU, Chang MK, Andriano KP. Mechanical properties of biodegradable polymers and composites proposed for internal fixation of bone. *J Appl Biomater* 1990;**1**:57–78. https://doi.org/10.1002/jab.770010109.
407. Tsai M-C, Hung K-C, Hung S-C, Hsu S. Evaluation of biodegradable elastic scaffolds made of anionic polyurethane for cartilage tissue engineering. *Colloids Surf B Biointerfaces* 2015;**125**:34–44. https://doi.org/10.1016/j.colsurfb.2014.11.003.
408. Girão AF, Semitela Â, Ramalho G, Completo A, Marques PAAP. Mimicking nature: Fabrication of 3D anisotropic electrospun polycaprolactone scaffolds for cartilage tissue engineering applications. *Compos B Eng* 2018;**154**:99–107. https://doi.org/10.1016/j.compositesb.2018.08.001.
409. Garrigues NW, Little D, Sanchez-Adams J, Ruch DS, Guilak F. Electrospun cartilage-derived matrix scaffolds for cartilage tissue engineering. *Journal of Biomedical Materials ResearchPart A* 2014;**102**:3998–4008. https://doi.org/10.1002/jbm.a.35068 [doi].
410. Larson BL, Yu SN, Park H, Estes BT, Moutos FT, Bloomquist CJ, *et al.* Chondrogenic, hypertrophic, and osteochondral differentiation of human mesenchymal stem cells on three-dimensionally woven scaffolds. *J Tissue Eng Regen Med* 2019;**13**:1453–65. https://doi.org/10.1002/TERM.2899.
411. Moutos FT, Freed LE, Guilak F. A biomimetic three-dimensional woven composite scaffold for functional tissue engineering of cartilage. *Nature Materials 2007 6:2* 2007;**6**:162–7. https://doi.org/10.1038/nmat1822.
412. Xia B, Kim D-H, Bansal S, Bae Y, Mauck RL, Heo S-J. Development of a decellularized meniscus matrix-based nanofibrous scaffold for meniscus tissue engineering. *Acta Biomater* 2021;**128**:175–85. https://doi.org/10.1016/j.actbio.2021.03.074.
413. Pillai MM, Gopinathan J, Senthil Kumar R, Sathish Kumar G, Shanthakumari S, Sahanand KS, *et al.* Tissue engineering of human knee meniscus using functionalized and reinforced silk-polyvinyl alcohol composite three-dimensional scaffolds: Understanding the in vitro and in vivo behavior. *J Biomed Mater Res A* 2018;**106**:1722–31. https://doi.org/10.1002/jbm.a.36372.
414. Jang CH, Koo Y, Kim G. ASC/chondrocyte-laden alginate hydrogel/PCL hybrid scaffold fabricated using 3D printing for auricle regeneration. *Carbohydr Polym* 2020;**248**:116776. https://doi.org/10.1016/j.carbpol.2020.116776.
415. Eviana Putri NR, Wang X, Chen Y, Li X, Kawazoe N, Chen G. Preparation of PLGA-collagen hybrid scaffolds with controlled pore structures for cartilage tissue engineering. *Progress*

in Natural Science: Materials International 2020;**30**:642–50. https://doi.org/10.1016/j.pnsc.2020.07.003.

416. Haaparanta A-M, Järvinen E, Cengiz IF, Ellä V, Kokkonen HT, Kiviranta I, *et al.* Preparation and characterization of collagen/PLA, chitosan/PLA, and collagen/chitosan/PLA hybrid scaffolds for cartilage tissue engineering. *J Mater Sci Mater Med* 2014;**25**:1129–36. https://doi.org/10.1007/s10856-013-5129-5.
417. Nogami M, Kimura T, Seki S, Matsui Y, Yoshida T, Koike-Soko C, *et al.* A Human Amnion-Derived Extracellular Matrix-Coated Cell-Free Scaffold for Cartilage Repair: In Vitro and In Vivo Studies. *Tissue Eng Part A* 2016;**22**:680–8. https://doi.org/10.1089/ten.TEA.2015.0285.
418. Lee JK, Hu JCY, Yamada S, Athanasiou KA. Initiation of Chondrocyte Self-Assembly Requires an Intact Cytoskeletal Network. *Tissue Eng Part A* 2016;**22**:318–25. https://doi.org/10.1089/ten.tea.2015.0491.
419. Huang AH, Motlekar NA, Stein A, Diamond SL, Shore EM, Mauck RL. High-throughput screening for modulators of mesenchymal stem cell chondrogenesis. *Ann Biomed Eng* 2008;**36**:1909–21. https://doi.org/10.1007/s10439-008-9562-4.
420. Frerker N, Karlsen TA, Lilledahl MB, Brorson S-H, Tibballs JE, Brinchmann JE. Scaffold-Free Engineering of Human Cartilage Implants. *Cartilage* 2021;**13**:1237S-1249S. https://doi.org/10.1177/19476035211007923.
421. Huang BJ, Hu JC, Athanasiou KA. Effects of passage number and post-expansion aggregate culture on tissue engineered, self-assembled neocartilage. *Acta Biomater* 2016;**43**:150–9. https://doi.org/10.1016/j.actbio.2016.07.044.
422. Kobayashi J, Kikuchi A, Aoyagi T, Okano T. Cell sheet tissue engineering: Cell sheet preparation, harvesting/manipulation, and transplantation. *J Biomed Mater Res A* 2019;**107**:955–67. https://doi.org/10.1002/jbm.a.36627.
423. Okano T, Yamada N, Okuhara M, Sakai H, Sakurai Y. Mechanism of cell detachment from temperature-modulated, hydrophilic-hydrophobic polymer surfaces. *Biomaterials* 1995;**16**:297–303. https://doi.org/10.1016/0142-9612(95)93257-e.
424. Kokubo M, Sato M, Yamato M, Mitani G, Uchiyama Y, Mochida J, *et al.* Characterization of layered chondrocyte sheets created in a co-culture system with synoviocytes in a hypoxic environment. *J Tissue Eng Regen Med* 2017;**11**:2885–94. https://doi.org/10.1002/term.2192.
425. Kondo M, Kameishi S, Kim K, Metzler NF, Maak TG, Hutchinson DT, *et al.* Preclinical safety and efficacy of juvenile cartilage-derived chondrocyte sheets for treating focal chondral injury. *Osteoarthritis Cartilage* 2021;**29**:S209–10. https://doi.org/10.1016/j.joca.2021.02.285.
426. Thorp H, Kim K, Bou-Ghannam S, Kondo M, Maak T, Grainger DW, *et al.* Enhancing chondrogenic potential via mesenchymal stem cell sheet multilayering. *Regen Ther* 2021;**18**:487–96. https://doi.org/10.1016/j.reth.2021.11.004.
427. Matsukura K, Kondo M, Metzler NF, Ford AJ, Maak TG, Hutchinson DT, *et al.* Regenerative Variability of Human Juvenile Chondrocyte Sheets From Different Cell Donors in an Athymic Rat Knee Chondral Defect Model. *Cartilage* 2024:19476035241277944. https://doi.org/10.1177/19476035241277946.
428. White JL, Walker NJ, Hu JC, White JL, Walker NJ, Hu JC, *et al.* A Comparison of Bone Marrow and Cord Blood Mesenchymal Stem Cells for Cartilage Self-Assembly 2018. https://doi.org/10.1089/ten.tea.2017.0424.
429. Dikina AD, Strobel HA, Lai BP, Rolle MW, Alsberg E. Engineered cartilaginous tubes for tracheal tissue replacement via self-assembly and fusion of human mesenchymal stem cell constructs. *Biomaterials* 2015;**52**:452–62. https://doi.org/10.1016/j.biomaterials.2015.01.073.
430. Hoburg A, Niemeyer P, Laute V, Zinser W, John T, Becher C, *et al.* Safety and Efficacy of Matrix-Associated Autologous Chondrocyte Implantation With Spheroids for Patellofemoral

or Tibiofemoral Defects: A 5-Year Follow-up of a Phase 2, Dose-Confirmation Trial. *Orthop J Sports Med* 2022;**10**:232596712110533. https://doi.org/10.1177/23259671211053380.
431. De Moor L, Beyls E, Declercq H. Scaffold Free Microtissue Formation for Enhanced Cartilage Repair. *Ann Biomed Eng* 2020;**48**:298–311. https://doi.org/10.1007/s10439-019-02348-4.
432. Oba T, Okamoto S, Ueno Y, Matsuo M, Tadokoro T, Kobayashi S, *et al.* In vitro elastic cartilage reconstruction using human auricular perichondrial chondroprogenitor cell-derived micro 3D spheroids. *J Tissue Eng* 2022;**13**:20417314221143484. https://doi.org/10.1177/20417314221143484.
433. Reutter S, Kern J, Jakob Y, Rotter N, Gvaramia D. Small spheroids for head and neck cartilage tissue engineering. *Sci Rep* 2024;**14**:32114. https://doi.org/10.1038/s41598-024-83847-w.
434. Gryadunova A, Kasamkattil J, Gay MHP, Dasen B, Pelttari K, Mironov V, *et al.* Nose to Spine: spheroids generated by human nasal chondrocytes for scaffold-free nucleus pulposus augmentation. *Acta Biomater* 2021;**134**:240–51. https://doi.org/10.1016/j.actbio.2021.07.064.
435. Kwon H, Paschos NK, Hu JC, Athanasiou K. Articular cartilage tissue engineering: the role of signaling molecules. *Cell Mol Life Sci* 2016;**73**:1173–94. https://doi.org/10.1007/s00018-015-2115-8.
436. Alexander TH, Sage AB, Schumacher BL, Sah RL, Watson D. Human serum for tissue engineering of human nasal septal cartilage. *Otolaryngol Head Neck Surg* 2006;**135**:397–403. https://doi.org/10.1016/j.otohns.2006.05.029.
437. Sato Y, Mera H, Takahashi D, Majima T, Iwasaki N, Wakitani S, *et al.* Synergistic effect of ascorbic acid and collagen addition on the increase in type 2 collagen accumulation in cartilage-like MSC sheet. *Cytotechnology* 2017;**69**:405–16. https://doi.org/10.1007/s10616-015-9924-3.
438. Chua KH, Aminuddin BS, Fuzina NH, Ruszymah BHI. Insulin-transferrin-selenium prevent human chondrocyte dedifferentiation and promote the formation of high quality tissue engineered human hyaline cartilage. *Eur Cell Mater* 2005;**9**:58–67; discussion 67. https://doi.org/10.22203/ecm.v009a08.
439. Thangavel P, Ramachandran B, Kannan R, Muthuvijayan V. Biomimetic hydrogel loaded with silk and l-proline for tissue engineering and wound healing applications. *J Biomed Mater Res B Appl Biomater* 2017;**105**:1401–8. https://doi.org/10.1002/jbm.b.33675.
440. Sekiya I, Koopman P, Tsuji K, Mertin S, Harley V, Yamada Y, *et al.* Dexamethasone enhances SOX9 expression in chondrocytes. *J Endocrinol* 2001;**169**:573–9. https://doi.org/10.1677/joe.0.1690573.
441. Zhong Y, Caplan AI, Welter JF, Baskaran H. Glucose Availability Affects Extracellular Matrix Synthesis During Chondrogenesis In Vitro. *Tissue Eng Part A* 2021;**27**:1321–32. https://doi.org/10.1089/ten.TEA.2020.0144.
442. Diaz-Romero J, Nesic D, Grogan SP, Heini P, Mainil-Varlet P. Immunophenotypic changes of human articular chondrocytes during monolayer culture reflect bona fide dedifferentiation rather than amplification of progenitor cells. *J Cell Physiol* 2008;**214**:75–83. https://doi.org/10.1002/jcp.21161.
443. Darling EM, Athanasiou KA. Rapid phenotypic changes in passaged articular chondrocyte subpopulations. *Journal of Orthopaedic Research* 2005;**23**:425–32. https://doi.org/10.1016/j.orthres.2004.08.008.
444. Oh SK, Kim HS, Park Y Bin, Seol HW, Kim YY, Cho MS, *et al.* Methods for expansion of human embryonic stem cells. *Stem Cells* 2005;**23**:605–9. https://doi.org/10.1634/stemcells.2004-0297.
445. Chan SW, Rizwan M, Yim EKF. Emerging Methods for Enhancing Pluripotent Stem Cell Expansion. *Front Cell Dev Biol* 2020;**8**:. https://doi.org/10.3389/fcell.2020.00070.

446. Lavon N, Zimerman M, Itskovitz-Eldor J. Scalable Expansion of Pluripotent Stem Cells. 2017. p. 23–37.
447. Miclau K, Hambright WS, Huard J, Stoddart MJ, Bahney CS. Cellular expansion of MSCs: Shifting the regenerative potential. *Aging Cell* 2023;**22**:e13759. https://doi.org/10.1111/acel.13759.
448. Hassan MNF Bin, Yazid MD, Yunus MHM, Chowdhury SR, Lokanathan Y, Idrus RBH, *et al.* Large-Scale Expansion of Human Mesenchymal Stem Cells. *Stem Cells Int* 2020;**2020**:9529465. https://doi.org/10.1155/2020/9529465.
449. Capelli C, Pedrini O, Valgardsdottir R, Da Roit F, Golay J, Introna M. Clinical grade expansion of MSCs. *Immunol Lett* 2015;**168**:222–7. https://doi.org/10.1016/j.imlet.2015.06.006.
450. Martin I, Vunjak-Novakovic G, Yang J, Langer R, Freed LE. Mammalian chondrocytes expanded in the presence of fibroblast growth factor 2 maintain the ability to differentiate and regenerate three-dimensional cartilaginous tissue. *Exp Cell Res* 1999;**253**:681–8. https://doi.org/10.1006/excr.1999.4708.
451. Hsieh-Bonassera ND, Wu I, Lin JK, Schumacher BL, Chen AC, Masuda K, *et al.* Expansion and redifferentiation of chondrocytes from osteoarthritic cartilage: cells for human cartilage tissue engineering. *Tissue Eng Part A* 2009;**15**:3513–23. https://doi.org/10.1089/ten.TEA.2008.0628.
452. Richmon JD, Sage AB, Shelton E, Schumacher BL, Sah RL, Watson D. Effect of growth factors on cell proliferation, matrix deposition, and morphology of human nasal septal chondrocytes cultured in monolayer. *Laryngoscope* 2005;**115**:1553–60. https://doi.org/10.1097/01.MLG.0000175541.31131.A5.
453. Adesida AB, Grady LM, Khan WS, Hardingham TE. The matrix-forming phenotype of cultured human meniscus cells is enhanced after culture with fibroblast growth factor 2 and is further stimulated by hypoxia. *Arthritis Res Ther* 2006;**8**:R61. https://doi.org/10.1186/ar1929.
454. Hwang NS, Varghese S, Elisseeff J. Derivation of Chondrogenically-Committed Cells from Human Embryonic Cells for Cartilage Tissue Regeneration. *PLoS One* 2008;**3**:e2498. https://doi.org/10.1371/journal.pone.0002498.
455. Nakagawa T, Lee SY, Reddi AH. Induction of chondrogenesis from human embryonic stem cells without embryoid body formation by bone morphogenetic protein 7 and transforming growth factor beta1. *Arthritis Rheum* 2009;**60**:3686–92. https://doi.org/10.1002/art.27229.
456. Kawata M, Mori D, Kanke K, Hojo H, Ohba S, Chung U-I, *et al.* Simple and Robust Differentiation of Human Pluripotent Stem Cells toward Chondrocytes by Two Small-Molecule Compounds. *Stem Cell Reports* 2019;**13**:530–44. https://doi.org/10.1016/j.stemcr.2019.07.012.
457. Peredo AP, Tsinman TK, Bonnevie ED, Jiang X, Smith HE, Gullbrand SE, *et al.* Developmental morphogens direct human induced pluripotent stem cells toward an annulus fibrosus-like cell phenotype. *JOR Spine* 2024;**7**:. https://doi.org/10.1002/jsp2.1313.
458. Uto S, Hikita A, Sakamoto T, Mori D, Yano F, Ohba S, *et al.* Ear Cartilage Reconstruction Combining Induced Pluripotent Stem Cell-Derived Cartilage and Three-Dimensional Shape-Memory Scaffold. *Tissue Eng Part A* 2021;**27**:604–17. https://doi.org/10.1089/ten.TEA.2020.0106.
459. Tallheden T, Karlsson C, Brunner A, van der Lee J, Hagg R, Tommasini R, *et al.* Gene expression during redifferentiation of human articular chondrocytes. *Osteoarthritis Cartilage* 2004;**12**:525–35. https://doi.org/10.1016/j.joca.2004.03.004.
460. Caron MMJ, Emans PJ, Coolsen MME, Voss L, Surtel DAM, Cremers A, *et al.* Redifferentiation of dedifferentiated human articular chondrocytes: comparison of 2D and 3D cultures. *Osteoarthritis Cartilage* 2012;**20**:1170–8. https://doi.org/10.1016/j.joca.2012.06.016.

461. Yang KGA, Saris DBF, Geuze RE, Helm YJM Van Der, Rijen MHP Van, Verbout AJ, *et al.* Impact of expansion and redifferentiation conditions on chondrogenic capacity of cultured chondrocytes. *Tissue Eng* 2006;**12**:2435–47. https://doi.org/10.1089/ten.2006.12.2435.
462. Hu X, Zhang W, Li X, Zhong D, Li Y, Li J, *et al.* Strategies to Modulate the Redifferentiation of Chondrocytes. *Front Bioeng Biotechnol* 2021;**9**:. https://doi.org/10.3389/fbioe.2021.764193.
463. Jakob M, Démarteau O, Schäfer D, Hintermann B, Dick W, Heberer M, *et al.* Specific growth factors during the expansion and redifferentiation of adult human articular chondrocytes enhance chondrogenesis and cartilaginous tissue formation in vitro. *J Cell Biochem* 2001;**81**:368–77. https://doi.org/10.1002/1097-4644(20010501)81:2<368::aid-jcb1051>3.0.co;2-j.
464. Weiss S, Hennig T, Bock R, Steck E, Richter W. Impact of growth factors and PTHrP on early and late chondrogenic differentiation of human mesenchymal stem cells. *J Cell Physiol* 2010;**223**:84–93. https://doi.org/10.1002/jcp.22013.
465. Sekiya I, Larson BL, Vuoristo JT, Reger RL, Prockop DJ. Comparison of effect of BMP-2, -4, and -6 on in vitro cartilage formation of human adult stem cells from bone marrow stroma. *Cell Tissue Res* 2005;**320**:269–76. https://doi.org/10.1007/s00441-004-1075-3.
466. Byers BA, Mauck RL, Chiang IE, Tuan RS. Transient exposure to transforming growth factor beta 3 under serum-free conditions enhances the biomechanical and biochemical maturation of tissue-engineered cartilage. *Tissue Eng Part A* 2008;**14**:1821–34. https://doi.org/10.1089/ten.tea.2007.0222.
467. Huang AH, Stein A, Tuan RS, Mauck RL. Transient exposure to transforming growth factor beta 3 improves the mechanical properties of mesenchymal stem cell-laden cartilage constructs in a density-dependent manner. *Tissue Eng Part A* 2009;**15**:3461–72. https://doi.org/10.1089/ten.TEA.2009.0198.
468. Murphy MK, Huey DJ, Hu JC, Athanasiou KA. TGF-beta1, GDF-5, and BMP-2 stimulation induces chondrogenesis in expanded human articular chondrocytes and marrow-derived stromal cells. *Stem Cells* 2015;**33**:762–73. https://doi.org/10.1002/stem.1890.
469. Domm C, Schünke M, Christesen K, Kurz B. Redifferentiation of dedifferentiated bovine articular chondrocytes in alginate culture under low oxygen tension. *Osteoarthritis Cartilage* 2002;**10**:13–22. https://doi.org/10.1053/joca.2001.0477.
470. Markway BD, Cho H, Johnstone B. Hypoxia promotes redifferentiation and suppresses markers of hypertrophy and degeneration in both healthy and osteoarthritic chondrocytes. *Arthritis Res Ther* 2013;**15**:R92. https://doi.org/10.1186/ar4272.
471. He A, Ye A, Song N, Liu N, Zhou G, Liu Y, *et al.* Phenotypic redifferentiation of dedifferentiated microtia chondrocytes through a three-dimensional chondrogenic culture system. *Am J Transl Res* 2020;**12**:2903–15.
472. Natoli R, Revell CM, Athanasiou K. Chondroitinase ABC Treatment Results in Increased Tensile Properties of Self-Assembled Tissue Engineered Articular Cartilage. *Tissue Eng Part A* 2009. https://doi.org/10.1089/ten.TEA.2008.0478.
473. Jin CZ, Cho J-H, Choi BH, Wang LM, Kim MS, Park SR, *et al.* The maturity of tissue-engineered cartilage in vitro affects the repairability for osteochondral defect. *Tissue Eng Part A* 2011;**17**:3057–65. https://doi.org/10.1089/ten.TEA.2010.0605.
474. Responte DJ, Arzi B, Natoli RM, Hu JC, Athanasiou KA. Mechanisms underlying the synergistic enhancement of self-assembled neocartilage treated with chondroitinase-ABC and TGF-beta1. *Biomaterials* 2012;**33**:3187–94. https://doi.org/10.1016/j.biomaterials.2012.01.028.

475. Makris EA, MacBarb RF, Paschos NK, Hu JC, Athanasiou KA. Combined use of chondroitinase-ABC, TGF-beta1, and collagen crosslinking agent lysyl oxidase to engineer functional neotissues for fibrocartilage repair. *Biomaterials* 2014;**35**:6787–96. https://doi.org/10.1016/j.biomaterials.2014.04.083.
476. Ng KW, Kugler LE, Doty SB, Ateshian GA, Hung CT. Scaffold degradation elevates the collagen content and dynamic compressive modulus in engineered articular cartilage. *Osteoarthritis Cartilage* 2009;**17**:220–7. https://doi.org/10.1016/j.joca.2008.06.013.
477. Salinas EY, Hu JC, Athanasiou K. A Guide for Using Mechanical Stimulation to Enhance Tissue-Engineered Articular Cartilage Properties. *Tissue Eng Part B Rev* 2018;**24**:345–58. https://doi.org/10.1089/ten.TEB.2018.0006.
478. Kowsari-Esfahan R, Jahanbakhsh A, Saidi MS, Bonakdar S. A microfabricated platform for the study of chondrogenesis under different compressive loads. *J Mech Behav Biomed Mater* 2018;**78**:404–13. https://doi.org/10.1016/j.jmbbm.2017.12.002.
479. Huang C-YC, Hagar KL, Frost LE, Sun Y, Cheung HS. Effects of cyclic compressive loading on chondrogenesis of rabbit bone-marrow derived mesenchymal stem cells. *Stem Cells* 2004;**22**:313–23. https://doi.org/10.1634/stemcells.22-3-313.
480. Bian L, Zhai DY, Zhang EC, Mauck RL, Burdick JA. Dynamic compressive loading enhances cartilage matrix synthesis and distribution and suppresses hypertrophy in hMSC-laden hyaluronic acid hydrogels. *Tissue Eng Part A* 2012;**18**:715–24. https://doi.org/10.1089/ten.TEA.2011.0455.
481. Nebelung S, Gavenis K, Lüring C, Zhou B, Mueller-Rath R, Stoffel M, *et al.* Simultaneous anabolic and catabolic responses of human chondrocytes seeded in collagen hydrogels to long-term continuous dynamic compression. *Ann Anat* 2012;**194**:351–8. https://doi.org/10.1016/j.aanat.2011.12.008.
482. Jeon JE, Schrobback K, Hutmacher DW, Klein TJ. Dynamic compression improves biosynthesis of human zonal chondrocytes from osteoarthritis patients. *Osteoarthritis Cartilage* 2012;**20**:906–15. https://doi.org/10.1016/j.joca.2012.04.019.
483. Aisenbrey EA, Bilousova G, Payne K, Bryant SJ. Dynamic mechanical loading and growth factors influence chondrogenesis of induced pluripotent mesenchymal progenitor cells in a cartilage-mimetic hydrogel. *Biomater Sci* 2019;**7**:5388–403. https://doi.org/10.1039/c9bm01081e.
484. Ge Y, Li Y, Wang Z, Li L, Teng H, Jiang Q. Effects of Mechanical Compression on Chondrogenesis of Human Synovium-Derived Mesenchymal Stem Cells in Agarose Hydrogel. *Front Bioeng Biotechnol* 2021;**9**:697281. https://doi.org/10.3389/fbioe.2021.697281.
485. Ballyns JJ, Bonassar LJ. Dynamic compressive loading of image-guided tissue engineered meniscal constructs. *J Biomech* 2011;**44**:509–16. https://doi.org/10.1016/j.jbiomech.2010.09.017.
486. Puetzer JL, Ballyns JJ, Bonassar LJ. The effect of the duration of mechanical stimulation and post-stimulation culture on the structure and properties of dynamically compressed tissue-engineered menisci. *Tissue Eng Part A* 2012;**18**:1365–75. https://doi.org/10.1089/ten.TEA.2011.0589.
487. Puetzer JL, Bonassar LJ. Physiologically Distributed Loading Patterns Drive the Formation of Zonally Organized Collagen Structures in Tissue-Engineered Meniscus. *Tissue Eng Part A* 2016;**22**:907–16. https://doi.org/10.1089/ten.TEA.2015.0519.
488. Puetzer JL, Koo E, Bonassar LJ. Induction of fiber alignment and mechanical anisotropy in tissue engineered menisci with mechanical anchoring. *J Biomech* 2015;**48**:1436–43. https://doi.org/10.1016/j.jbiomech.2015.02.033.

489. Huey DJ, Athanasiou KA. Tension-compression loading with chemical stimulation results in additive increases to functional properties of anatomic meniscal constructs. *PLoS One* 2011;**6**:e27857. https://doi.org/10.1371/journal.pone.0027857.
490. Chung C, Erickson IE, Mauck RL, Burdick JA. Differential behavior of auricular and articular chondrocytes in hyaluronic acid hydrogels. *Tissue EngineeringPart A* 2008;**14**:1121–31. https://doi.org/10.1089/tea.2007.0291.
491. Lee JK, Huwe LW, Paschos N, Aryaei A, Gegg CA, Hu JC, *et al.* Tension stimulation drives tissue formation in scaffold-free systems. *Nat Mater* 2017;**16**:864–73. https://doi.org/10.1038/nmat4917.
492. Baker BM, Shah RP, Huang AH, Mauck RL. Dynamic tensile loading improves the functional properties of mesenchymal stem cell-laden nanofiber-based fibrocartilage. *Tissue Eng Part A* 2011;**17**:1445–55. https://doi.org/10.1089/ten.tea.2010.0535.
493. Puetzer JL, Ma T, Sallent I, Gelmi A, Stevens MM. Driving Hierarchical Collagen Fiber Formation for Functional Tendon, Ligament, and Meniscus Replacement. *Biomaterials* 2021;**269**:120527. https://doi.org/10.1016/j.biomaterials.2020.120527.
494. Elder BD, Athanasiou KA. Synergistic and Additive Effects of Hydrostatic Pressure and Growth Factors on Tissue Formation. *PLoS One* 2008;**3**:e2341. https://doi.org/10.1371/journal.pone.0002341.
495. Elder BD, Athanasiou KA. Effects of temporal hydrostatic pressure on tissue-engineered bovine articular cartilage constructs. *Tissue Eng Part A* 2009;**15**:1151–8. https://doi.org/10.1089/ten.tea.2008.0200.
496. Tatsumura M, Sakane M, Ochiai N, Mizuno S. Off-loading of cyclic hydrostatic pressure promotes production of extracellular matrix by chondrocytes. *Cells Tissues Organs* 2013;**198**:405–13. https://doi.org/10.1159/000360156.
497. Hodder E, Guppy F, Covill D, Bush P. The effect of hydrostatic pressure on proteoglycan production in articular cartilage in vitro: a meta-analysis. *Osteoarthritis Cartilage* 2020;**28**:1007–19. https://doi.org/10.1016/j.joca.2020.03.021.
498. Almarza AJ, Athanasiou KA. Effects of hydrostatic pressure on TMJ disc cells. *Tissue Eng* 2006;**12**:1285–94. https://doi.org/10.1089/ten.2006.12.1285.
499. Gunja NJ, Uthamanthil RK, Athanasiou KA. Effects of TGF-β1 and hydrostatic pressure on meniscus cell-seeded scaffolds. *Biomaterials* 2009;**30**:565–73. https://doi.org/10.1016/j.biomaterials.2008.10.007.
500. Gharravi AM, Orazizadeh M, Ansari-Asl K, Banoni S, Izadi S, Hashemitabar M. Design and fabrication of anatomical bioreactor systems containing alginate scaffolds for cartilage tissue engineering. *Avicenna J Med Biotechnol* 2012;**4**:65–74.
501. Gemmiti C V, Guldberg RE. Fluid flow increases type II collagen deposition and tensile mechanical properties in bioreactor-grown tissue-engineered cartilage. *Tissue Eng* 2006;**12**:469–79. https://doi.org/10.1089/ten.2006.12.469.
502. Gemmiti C V, Guldberg RE. Shear stress magnitude and duration modulates matrix composition and tensile mechanical properties in engineered cartilaginous tissue. *Biotechnol Bioeng* 2009;**104**:809–20. https://doi.org/10.1002/bit.22440.
503. Salinas EY, Aryaei A, Paschos N, Berson E, Kwon H, Hu JC, *et al.* Shear stress induced by fluid flow produces improvements in tissue-engineered cartilage. *Biofabrication* 2020;**12**:045010. https://doi.org/10.1088/1758-5090/ABA412.
504. Grad S, Loparic M, Peter R, Stolz M, Aebi U, Alini M. Sliding motion modulates stiffness and friction coefficient at the surface of tissue engineered cartilage. *Osteoarthritis Cartilage* 2012;**20**:288–95. https://doi.org/10.1016/j.joca.2011.12.010.

505. Detamore MS, Athanasiou KA. Use of a rotating bioreactor toward tissue engineering the temporomandibular joint disc. *Tissue Eng* 2005;**11**:1188–97. https://doi.org/10.1089/ten.2005.11.1188.
506. Li S, Oreffo ROC, Sengers BG, Tare RS. The effect of oxygen tension on human articular chondrocyte matrix synthesis: Integration of experimental and computational approaches. *Biotechnol Bioeng* 2014;**111**:1876–85. https://doi.org/10.1002/bit.25241.
507. Cissell DD, Link JM, Hu JC, Athanasiou KA. A Modified Hydroxyproline Assay Based on Hydrochloric Acid in Ehrlich's Solution Accurately Measures Tissue Collagen Content. *Tissue Eng Part C Methods* 2017;**23**:243–50. https://doi.org/10.1089/ten.tec.2017.0018.
508. Junqueira LCU, Bignolas G, Brentani RR. A simple and sensitive method for the quantitative estimation of collagen. *Anal Biochem* 1979;**94**:96–9. https://doi.org/10.1016/0003-2697(79)90795-4.
509. Yasmin H, Kabashima T, Rahman MS, Shibata T, Kai M. Amplified and selective assay of collagens by enzymatic and fluorescent reactions. *Sci Rep* 2014;**4**:4950. https://doi.org/10.1038/srep04950.
510. Farndale RW, Buttle DJ, Barrett AJ. Improved quantitation and discrimination of sulphated glycosaminoglycans by use of dimethylmethylene blue. *Biochimica et Biophysica Acta (BBA) - General Subjects* 1986;**883**:173–7. https://doi.org/10.1016/0304-4165(86)90306-5.
511. Almarza AJ, Bean AC, Baggett LS, Athanasiou KA. Biochemical analysis of the porcine temporomandibular joint disc. *British Journal of Oral and Maxillofacial Surgery* 2006;**44**:124–8. https://doi.org/10.1016/j.bjoms.2005.05.002.
512. Patel JM, Wise BC, Bonnevie ED, Mauck RL. A Systematic Review and Guide to Mechanical Testing for Articular Cartilage Tissue Engineering. *Tissue Eng Part C Methods* 2019;**25**:593–608. https://doi.org/10.1089/ten.tec.2019.0116.
513. Link JM, Hu JC, Athanasiou KA. Chondroitinase ABC Enhances Integration of Self-Assembled Articular Cartilage, but Its Dosage Needs to Be Moderated Based on Neocartilage Maturity. *Cartilage* 2020:1947603520918653. https://doi.org/10.1177/1947603520918653.
514. Nordberg RC, Wen DH, Wang D, Hu JC, Athanasiou KA. Challenges and recent advances in engineering the osteochondral interface. *Curr Opin Biomed Eng* 2024;**31**:100546. https://doi.org/10.1016/j.cobme.2024.100546.
515. Mow VC, Kuei SC, Lai WM, Armstrong CG. Biphasic creep and stress relaxation of articular cartilage in compression? Theory and experiments. *J Biomech Eng* 1980;**102**:73–84. https://doi.org/10.1115/1.3138202.
516. Athanasiou KA, Natoli RM. *Introduction to Continuum Biomechanics.* Cham: Springer International Publishing; 2009.
517. Elliott AD. Confocal Microscopy: Principles and Modern Practices. *Curr Protoc Cytom* 2020;**92**:. https://doi.org/10.1002/cpcy.68.
518. Miranda K, Girard-Dias W, Attias M, de Souza W, Ramos I. Three dimensional reconstruction by electron microscopy in the life sciences: An introduction for cell and tissue biologists. *Mol Reprod Dev* 2015:530–47. https://doi.org/10.1002/mrd.22455.
519. Basso P, Caravà E, Protasoni M, Reguzzoni M, Raspanti M. *The synovial surface of the articular cartilage.* vol. 64. 2020.
520. Brown WE, Huey DJ, Hu JC, Athanasiou KA. Functional self-assembled neocartilage as part of a biphasic osteochondral construct. *PLoS One* 2018;**13**:e0195261. https://doi.org/10.1371/journal.pone.0195261.
521. Ramos R, Pham KT, Prince RC, Leiser-Miller LB, Prasad MS, Wang X, *et al. Superstable lipid vacuoles endow cartilage with its shape and biomechanics.* n.d.
522. Bergholt MS, Serio A, Albro MB. Raman Spectroscopy: Guiding Light for the Extracellular Matrix. *Front Bioeng Biotechnol* 2019. https://doi.org/10.3389/fbioe.2019.00303.

523. Castro NJ, Babakhanova G, Hu J, Athanasiou KA. Nondestructive testing of native and tissue-engineered medical products: adding numbers to pictures. *Trends Biotechnol* 2022;**40**:194–209. https://doi.org/10.1016/J.TIBTECH.2021.06.009.

524. Bergholt MS, St-Pierre J-P, Offeddu GS, Parmar PA, Albro MB, Puetzer JL, *et al.* Raman Spectroscopy Reveals New Insights into the Zonal Organization of Native and Tissue-Engineered Articular Cartilage. *ACS Cent Sci* 2016;**2**:885–95. https://doi.org/10.1021/acscentsci.6b00222.

525. Cornett DS, Reyzer ML, Chaurand P, Caprioli RM. MALDI imaging mass spectrometry: Molecular snapshots of biochemical systems. *Nat Methods* 2007;**4**:828–33. https://doi.org/10.1038/nmeth1094.

526. Graham DJ, Gamble LJ. Back to the basics of time-of-flight secondary ion mass spectrometry of bio-related samples. I. Instrumentation and data collection. *Biointerphases* 2023;**18**:021201. https://doi.org/10.1116/6.0002477.

527. Nordberg RC, Zhang J, Griffith EH, Frank MW, Starly B, Loboa EG. Electrical Cell-Substrate Impedance Spectroscopy Can Monitor Age-Grouped Human Adipose Stem Cell Variability During Osteogenic Differentiation. *Stem Cells Transl Med* 2017;**6**:502–11. https://doi.org/10.5966/sctm.2015-0404 [doi].

528. Torrado B, Pannunzio B, Malacrida L, Digman MA. Fluorescence lifetime imaging microscopy. *Nature Reviews Methods Primers* 2024;**4**:80. https://doi.org/10.1038/s43586-024-00358-8.

529. Haudenschild AK, Sherlock BE, Zhou X, Sheaff CS, Hu JC, Leach JK, *et al.* Non-destructive, continuous monitoring of biochemical, mechanical, and structural maturation in engineered tissue. *Sci Rep* 2022;**12**:. https://doi.org/10.1038/s41598-022-18702-x.

530. Krieg M, Fläschner G, Alsteens D, Gaub BM, Roos WH, Wuite GJL, *et al.* Atomic force microscopy-based mechanobiology. *Nature Reviews Physics* 2019:41–57. https://doi.org/10.1038/s42254-018-0001-7.

531. Li X, Wang C-Y. From bulk, single-cell to spatial RNA sequencing. *Int J Oral Sci* 2021;**13**:36. https://doi.org/10.1038/s41368-021-00146-0.

532. Symons JL, Cho KJ, Chang JT, Du G, Waxham MN, Hancock JF, *et al.* Lipidomic atlas of mammalian cell membranes reveals hierarchical variation induced by culture conditions, subcellular membranes, and cell lineages. *Soft Matter* 2021;**17**:288–97. https://doi.org/10.1039/d0sm00404a.

533. Hemstrom W, Grummer JA, Luikart G, Christie MR. Next-generation data filtering in the genomics era. *Nat Rev Genet* 2024. https://doi.org/10.1038/s41576-024-00738-6.

534. Liu X, Locasale JW. Metabolomics: A Primer. *Trends Biochem Sci* 2017;**42**:274–84. https://doi.org/10.1016/j.tibs.2017.01.004.

535. MacBarb RF, Chen AL, Hu JC, Athanasiou KA. Engineering functional anisotropy in fibrocartilage neotissues. *Biomaterials* 2013;**34**:9980–9. https://doi.org/10.1016/j.biomaterials.2013.09.026.

536. Elder BD, Athanasiou KA. Systematic assessment of growth factor treatment on biochemical and biomechanical properties of engineered articular cartilage constructs. *Osteoarthritis Cartilage* 2009;**17**:114–23. https://doi.org/10.1016/j.joca.2008.05.006.

537. US Food and Drug Administration, FDA/CBER. *Guidance for Industry: Preclinical Assessment of Investigational Cellular and Gene Therapy Products.* 2013.

538. US Food and Drug Administration. Guidance for Industry: Preparation of IDEs and INDs for Products Intended to Repair or Replace Knee Cartilage 2011.

539. Cook JL, Hung CT, Kuroki K, Stoker AM, Cook CR, Pfeiffer FM, *et al.* Animal models of cartilage repair. *Bone Joint Res* 2014;**3**:89–94. https://doi.org/10.1302/2046-3758.34.2000238.

540. Donahue RP, Nordberg RC, Bielajew BJ, Hu JC, Athanasiou KA. The Effect of Neonatal, Juvenile, and Adult Donors on Rejuvenated Neocartilage Functional Properties. *Tissue Eng Part A* 2022;**28**:383–93. https://doi.org/10.1089/TEN.TEA.2021.0167.
541. FDA. Guidance for Industry Product Development Under the Animal Rule. *Food and Drug Administration* 2015:54.
542. Food and Drug Administration, Food, Administration D, Food and Drug Administration. *Points to Consider in the Characterization of Cell Lines Used to Produce Biologicals.* 1993.
543. FDA. Guidance for Industry Cell Characterization and Qualification of Cell Substrates and Other Biological Materials Used in the Production of Viral Vaccines for Infectious Disease Indications 2010:1–50.
544. US Food and Drug Administration. Guidance for FDA Reviewers and Sponsors: Content and Review of Chemistry, Manufacturing, and Control (CMC) Information for Human Somatic Cell Therapy Investigational New Drug Applications (INDs) 2008.
545. Chu CR, Szczodry M, Bruno S. Animal models for cartilage regeneration and repair. *Tissue Eng Part B Rev* 2010;**16**:105–15. https://doi.org/10.1089/ten.TEB.2009.0452.
546. Vacanti CA, Cima LG, Ratkowski D, Upton J, Vacanti JP. Tissue Engineered Growth of New Cartilage in the Shape of a Human Ear Using Synthetic Polymers Seeded with Chondrocytes. *MRS Proceedings* 1991;**252**:367. https://doi.org/10.1557/PROC-252-367.
547. Reiffel AJ, Kafka C, Hernandez KA, Popa S, Perez JL, Zhou S, *et al.* High-fidelity tissue engineering of patient-specific auricles for reconstruction of pediatric microtia and other auricular deformities. *PLoS One* 2013;**8**:e56506. https://doi.org/10.1371/journal.pone.0056506.
548. Shapiro F, Koide S, Glimcher MJ. Cell origin and differentiation in the repair of full-thickness defects of articular cartilage. *JBJS* 1993;**75**:.
549. ASTM International. Standard Guide for in vivo Assessment of Implantable Devices Intended to Repair or Regenerate Articular Cartilage F2451 – 05 (Withdrawn 2019). *ASTM International* 2010.
550. Hurtig MB, Buschmann MD, Fortier LA, Hoemann CD, Hunziker EB, Jurvelin JS, *et al.* Preclinical Studies for Cartilage Repair: Recommendations from the International Cartilage Repair Society. *Cartilage* 2011;**2**:137–52. https://doi.org/10.1177/1947603511401905.
551. Estes BT, Enomoto M, Moutos FT, Carson MA, Toth JM, Eggert P, *et al.* Biological resurfacing in a canine model of hip osteoarthritis. *Sci Adv* 2021;**7**:5918–33. 10.1126/SCIADV.ABI5918/SUPPL_FILE/SCIADV.ABI5918_SM.PDF.
552. ASTM. Standard Guide for Pre-clinical in vivo Evaluation of Spinal Fusion 1. *Astm* 2015. https://doi.org/10.1520/F2884.
553. Lunney JK, Goor A Van, Walker KE, Hailstock T, Franklin J, Dai C. Importance of the pig as a human biomedical model. *Sci Transl Med* 2021;**13**:. https://doi.org/10.1126/SCITRANSLMED.ABD5758.
554. Cone SG, Warren PB, Fisher MB. Rise of the Pigs: Utilization of the Porcine Model to Study Musculoskeletal Biomechanics and Tissue Engineering During Skeletal Growth. *Tissue Eng Part C Methods* 2017;**23**:763–80. https://doi.org/10.1089/ten.TEC.2017.0227.
555. Nordberg RC, Otarola GA, Wang D, Hu JC, Athanasiou KA. Navigating regulatory pathways for translation of biologic cartilage repair products. *Sci Transl Med* 2024;**14**:eabp8163. https://doi.org/10.1126/scitranslmed.abp8163.
556. Chen D, Wu JY, Kennedy KM, Yeager K, Bernhard JC, Ng JJ, *et al.* Tissue engineered autologous cartilage-bone grafts for temporomandibular joint regeneration. *Sci Transl Med* 2020;**12**:eabb6683. https://doi.org/10.1126/scitranslmed.abb6683.
557. Otsuki S, Nakagawa K, Murakami T, Sezaki S, Sato H, Suzuki M, *et al.* Evaluation of Meniscal Regeneration in a Mini Pig Model Treated With a Novel Polyglycolic Acid Meniscal Scaffold.

American Journal of Sports Medicine 2019;**47**:1804–15. https://doi.org/10.1177/0363546519850578.

558. FDA. *Guidance for Industry and/or FDA Reviewers/Staff Guidance Document for the Preparation of IDEs for Spinal Systems.* 2000.
559. Wang D, Cubberly M, Brown WE, Kwon H, Hu JC, Athanasiou KA. Diagnostic Arthroscopy of the Minipig Stifle (Knee) for Translational Large Animal Research. *Arthrosc Tech* 2021;**10**:e297–301. https://doi.org/10.1016/j.eats.2020.10.005.
560. Jackson DW, Lalor PA, Aberman HM, Simon TM. Spontaneous repair of full-thickness defects of articular cartilage in a goat model. *J Bone Joint Surg* 2001;**23-A**:53–64.
561. von Rechenberg B, Akens MK, Nadler D, Bittmann P, Zlinszky K, Kutter A, *et al.* Changes in subchondral bone in cartilage resurfacing - An experimental study in sheep using different types of osteochondral grafts. *Osteoarthritis Cartilage* 2003;**11**:265–77. https://doi.org/10.1016/S1063-4584(03)00006-2.
562. US Food and Drug Administration. MACI ® (autologous cultured chondrocytes on porcine collagen membrane) Cellular sheet for autologous implantation 2021.
563. Kol A, Arzi B, Athanasiou KA, Farmer DL, Nolta JA, Rebhun RB, *et al.* Companion animals: Translational scientist's new best friends. *Sci Transl Med* 2015;**7**:308ps21–308ps21. https://doi.org/10.1126/scitranslmed.aaa9116.
564. Modesto K. A.; Schaer T. P. RB; M. Arthroscopy of the normal cadaveric ovine femorotibial joint: a systematic approach to the cranial and caudal compartments. *Veterinary and Comparative Orthopaedics and Traumatology* 2014;**27**:387–94. https://doi.org/10.3415/VCOT-14-03-0039.
565. Frisbie DD, Barrett MF, McIlwraith CW, Ullmer J. Diagnostic stifle joint arthroscopy using a needle arthroscope in standing horses. *Veterinary Surgery* 2014;**43**:12–8. https://doi.org/10.1111/j.1532-950X.2013.12068.x.
566. Oei Marius C.; Müller-Lutz Anja; Schleich Christoph; Miese Falk R. EHG; W. Cartilage Imaging: Techniques and Developments. *Semin Musculoskelet Radiol* 2018;**22**:245–60. https://doi.org/10.1055/s-0038-1639471.
567. Kellgren JH, Lawrence JS. Radiological Assessment of Osteo-Arthrosis. *Ann Rheum Dis* 1957;**16**:494–502. https://doi.org/10.1136/ard.16.4.494.
568. Athanasiou KA. *Articular Cartilage.* CRC Press; 2017.
569. Marlovits S, Striessnig G, Resinger CT, Aldrian SM, Vecsei V, Imhof H, *et al.* Definition of pertinent parameters for the evaluation of articular cartilage repair tissue with high-resolution magnetic resonance imaging. *Eur J Radiol* 2004;**52**:310–9. https://doi.org/10.1016/j.ejrad.2004.03.014.
570. Marlovits S, Singer P, Zeller P, Mandl I, Haller J, Trattnig S. Magnetic resonance observation of cartilage repair tissue (MOCART) for the evaluation of autologous chondrocyte transplantation: Determination of interobserver variability and correlation to clinical outcome after 2 years. *Eur J Radiol* 2006;**57**:16–23. https://doi.org/10.1016/j.ejrad.2005.08.007.
571. Fisher MB, Belkin NS, Milby AH, Henning EA, Bostrom M, Kim M, *et al.* Cartilage Repair and Subchondral Bone Remodeling in Response to Focal Lesions in a Mini-Pig Model: Implications for Tissue Engineering. *Tissue Eng Part A* 2015;**21**:850–60. https://doi.org/10.1089/ten.tea.2014.0384.
572. Nordberg RC, Huebner P, Schuchard KG, Mellor LF, Shirwaiker RA, Loboa EG, *et al.* The evaluation of a multiphasic 3D-bioplotted scaffold seeded with adipose derived stem cells to repair osteochondral defects in a porcine model. *J Biomed Mater Res B Appl Biomater* 2021;**109**:2246–58. https://doi.org/10.1002/JBM.B.34886.
573. Deuis JR, Dvorakova LS, Vetter I. Methods used to evaluate pain behaviors in rodents. *Front Mol Neurosci* 2017. https://doi.org/10.3389/fnmol.2017.00284.

574. Turner P V., Pang DSJ, Lofgren JLS. A Review of Pain Assessment Methods in Laboratory Rodents. *Comp Med* 2019:451–67. https://doi.org/10.30802/AALAS-CM-19-000042.
575. Tomacheuski RM, Monteiro BP, Evangelista MC, Luna SPL, Steagall PV. Measurement properties of pain scoring instruments in farm animals: A systematic review using the COSMIN checklist. *PLoS One* 2023;**18**:. https://doi.org/10.1371/journal.pone.0280830.
576. Cimino Brown D, Boston RC, Coyne JC, Farrar JT. *Ability of the Canine Brief Pain Inventory to detect response to treatment in dogs with osteoarthritis.* n.d.
577. Sveum JW, Mishra RR, Marti TL, Jones JM, Hellenbrand DJ, Hanna AS. Gait analysis in swine, sheep, and goats after neurologic injury: a literature review. *Neural Regen Res* 2023:1917–24. https://doi.org/10.4103/1673-5374.367839.
578. Netzley AH, Hunt RD, Franco-Arellano J, Arnold N, Vazquez AI, Munoz KA, *et al.* Multimodal characterization of Yucatan minipig behavior and physiology through maturation. *Sci Rep* 2021;**11**:. https://doi.org/10.1038/s41598-021-00782-w.
579. Pritzker KPH, Gay S, Jimenez SA, Ostergaard K, Pelletier J-P, Revell PA, *et al.* Osteoarthritis cartilage histopathology: grading and staging. *Osteoarthritis Cartilage* 2006;**14**:13–29. https://doi.org/10.1016/j.joca.2005.07.014.
580. O'Driscoll SW, Marx RG, Beaton DE, Miura Y, Gallay SH, Fitzsimmons JS. Validation of a Simple Histological-Histochemical Cartilage Scoring System. *Tissue Eng* 2001;**7**:313–20. https://doi.org/10.1089/10763270152044170.
581. Mainil-Varlet P, Van Damme B, Nesic D, Knutsen G, Kandel R, Roberts S. A new histology scoring system for the assessment of the quality of human cartilage repair: ICRS II. *American Journal of Sports Medicine* 2010;**38**:880–90. https://doi.org/10.1177/0363546509359068.
582. Takahashi T, Donahue R, Nordberg R, Hu J, Currall S, Athanasiou K. Commercialization of regenerative medicine therapies. *Nature Reviews Bioengineering* 2023.
583. Wood JJ, Malek MA, Frassica FJ, Polder JA, Mohan AK, Bloom ET, *et al.* Autologous cultured chondrocytes: adverse events reported to the United States Food and Drug Administration. *J Bone Joint Surg Am* 2006;**88**:503–7. https://doi.org/10.2106/JBJS.E.00103.
584. Office of Tissues and Advanced Therapies. *Foreign Regulatory Seminar*. Food and Drug Administration. 2020.
585. Food and Drug Administration. *21 CFR Part 1271- Human Cells, Tissues, and Cellular and Tissue-Based Products.* https://Www.Ecfr.Gov/Current/Title-21/Part-1271. n.d.
586. Fda, Cber. *Guidance for Industry: Considerations for the Design of Early-Phase Clinical Trials of Cellular and Gene Therapy Products.* 2015.
587. Fda, Cber. *Manufacturing Changes and Comparability for Human Cellular and Gene Therapy Products; Draft Guidance for Industry.* n.d.
588. Pope, Pamela. *Potency Assurance for Cellular and Gene Therapy Products Draft Guidance for Industry.* 2009.
589. Fda, Cber. *Considerations for the Use of Human-and Animal-Derived Materials in the Manufacture of Cellular and Gene Therapy and Tissue-Engineered Medical Products Draft Guidance for Industry.* n.d.
590. Food and Drug Administration. *Safety Testing of Human Allogeneic Cells Expanded for Use in Cell-Based Medical Products.* https://Www.Fda.Gov/Regulatory-Information/Search-Fda-Guidance-Documents/Safety-Testing-Human-Allogeneic-Cells-Expanded-Use-Cell-Based-Medical-Products. n.d.
591. Fda, Cber. *Long Term Follow-Up After Administration of Human Gene Therapy Products; Guidance for Industry.* 2020.
592. Fda, Cber. *Chemistry, Manufacturing, and Control (CMC) Information for Human Gene Therapy Investigational New Drug Applications (INDs) Guidance for Industry.* 2020.

593. Fda, Cber. *Human Gene Therapy Products Incorporating Human Genome Editing; Guidance for Industry*. n.d.
594. International Organization for Standardization. *ISO 13485:2016*. https://Www.Iso.Org/Standard/59752.Html. n.d.
595. International Organization for Standardization. *ISO 11737–1:2018*. https://Www.Iso.Org/Standard/66451.Html. n.d.
596. International Organization for Standardization. *ISO 14644–1:2015*. https://Www.Iso.Org/Standard/53394.Html. n.d.
597. International Organization for Standardization. *ISO 14971:2019*. https://Www.Iso.Org/Standard/72704.Html. n.d.
598. International Organization for Standardization. ISO 20399:2022. https://wwwI.soOrg/Standard/79399Html n.d.
599. International Organization for Standardization. *ISO 10993–1:2018*. https://Www.Iso.Org/Standard/68936.Html. n.d.
600. International Organization for Standardization. *21973:2020*. https://Www.Iso.Org/Standard/72326.Html. n.d.
601. International Council for Harmonisation. *Quality Guidelines*. https://Ich.Org/Page/Quality-Guidelines#5-2. n.d.
602. ASTM. *ASTM F2027-16: Standard Guide for Characterization and Testing of Raw or Starting Materials for Tissue-Engineered Medical Products*. https://Store.Astm.Org/F2027-16.Html. n.d.
603. ASTM. *ASTM F2451-05: Standard Guide for in vivo Assessment of Implantable Devices Intended to Repair or Regenerate Articular Cartilage*. https://Store.Astm.Org/F2451-05.Html. n.d.
604. Food and Drug Administration. *21st Century Cures Act*. https://Www.Fda.Gov/Regulatory-Information/Selected-Amendments-Fdc-Act/21st-Century-Cures-Act. n.d.
605. Food and Drug Administration. *Regenerative Medicine Advanced Therapy Designation*. https://Www.Fda.Gov/Vaccines-Blood-Biologics/Cellular-Gene-Therapy-Products/Regenerative-Medicine-Advanced-Therapy-Designation. n.d.
606. *42 USC Ch. 6A: PUBLIC HEALTH SERVICE.*. https://Uscode.House.Gov/View.Xhtml?Path=/Prelim@title42/Chapter6A&edition=prelim. n.d.
607. Code of Federal Regulations. *21 CFR 1271.3 -- How does FDA define important terms in this part?*. https://Www.Ecfr.Gov/Current/Title-21/Part-1271/Section-1271.3. n.d.
608. Code of Federal Regulations. *21 CFR Part 312 -- Investigational New Drug Application*. https://Www.Ecfr.Gov/Current/Title-21/Chapter-I/Subchapter-L/Part-1271/Subpart-A/Section-1271.3. n.d.
609. Code of Federal Regulation. *21 CFR 210.2 -- Applicability of current good manufacturing practice regulations*. https://Www.Ecfr.Gov/Current/Title-21/Part-210/Section-210.2. n.d.
610. Wang AS, Nagelli C V, Lamba A, Saris DBF, Krych AJ, Hevesi M. Minimum 10-Year Outcomes of Matrix-Induced Autologous Chondrocyte Implantation in the Knee: A Systematic Review. *Am J Sports Med* 2024;**52**:2407–14. https://doi.org/10.1177/03635465231205309.
611. Andriolo L, Merli G, Filardo G, Marcacci M, Kon E. Failure of Autologous Chondrocyte Implantation. *Sports Med Arthrosc Rev* 2017;**25**:10–8. https://doi.org/10.1097/JSA.0000000000000137.
612. Kunze KN, Persaud S, Briano J, Rodeo SA, Warren RF, Wickiewicz TL, *et al.* Outcomes of Revision Cartilage Restoration Surgery for Failed Primary Treatment of Chondral or Osteochondral Defects of the Knee: A Systematic Review. *Am J Sports Med* 2025;**53**:1515–23. https://doi.org/10.1177/03635465241260271.

613. Beck A, Wood D, Vertullo CJ, Ebert J, Janes G, Sullivan M, *et al.* Morphological Assessment of MACI Grafts in Patients with Revision Surgery and Total Joint Arthroplasty. *Cartilage* 2021;**13**:526S-539S. https://doi.org/10.1177/1947603519890754.
614. Biosolution Co. Ltd. *NCT04744402: Phase 2 Clinical Trial of CartiLife® in the United States.* 2023.
615. 3DBio Therapeutics. *NCT04399239: AuriNovo for Auricular Reconstruction.* 2021.
616. Olivos-Meza A, Pérez Jiménez FJ, Granados-Montiel J, Landa-Solís C, Cortés González S, Jiménez Aroche CA, *et al.* First Clinical Application of Polyurethane Meniscal Scaffolds with Mesenchymal Stem Cells and Assessment of Cartilage Quality with T2 Mapping at 12 Months. *Cartilage* 2021;**13**:197S-207S. https://doi.org/10.1177/1947603519852415.
617. DiscGenics Inc. *NCT03955315: A Randomized, Double Blind, Sham-Controlled, Multi-Center Phase I / II Clinical Study to Assess the Safety and Efficacy of IDCT in Patients With Lumbar Degenerative Disc Disease.* 2022.
618. DiscGenics Inc. *NCT03347708: A Prospective, Randomized, Double-Blinded, Vehicle- and Placebo-Controlled, Multicenter Study to Evaluate the Safety and Preliminary Efficacy of IDCT in Subjects With Single-Level, Symptomatic Lumbar Intervertebral Disc Degeneration.* 2023.
619. Gornet MF, Beall DP, Davis TT, Coric D, LaBagnara M, Krull A, *et al.* Allogeneic Disc Progenitor Cells Safely Increase Disc Volume and Improve Pain, Disability, and Quality of Life in Patients With Lumbar Disc Degeneration-Results of an FDA-Approved Biologic Therapy Randomized Clinical Trial. *Int J Spine Surg* 2024;**18**:237–48. https://doi.org/10.14444/8609.
620. Mesoblast Ltd. *NCT02412735: A Prospective, Multicenter, Randomized, Double-Blind, Placebo-Controlled Study to Evaluate the Efficacy and Safety of a Single Injection of Rexlemestrocel-L Alone or Combined With Hyaluronic Acid (HA) in Subjects With Chronic Low Back Pain.* 2022.
621. Mesoblast Ltd. *NCT06325566: A Prospective, Multicenter, Randomized, Double-Blind, Controlled Study to Evaluate the Efficacy and Safety of a Single Injection of Rexlemestrocel-L Combined With HA in Subjects With Moderate to Severe Chronic Low Back Pain.* 2025.
622. Murphy MK, Masters TE, Hu JC, Athanasiou KA. Engineering a fibrocartilage spectrum through modulation of aggregate redifferentiation. *Cell Transplant* 2015;**24**:235–45.

MIX
Papier aus verantwortungsvollen Quellen
Paper from responsible sources
FSC® C105338

If you have any concerns about our products,
you can contact us on
ProductSafety@springernature.com

In case Publisher is established outside the EU,
the EU authorized representative is:
Springer Nature Customer Service Center GmbH
Europaplatz 3, 69115 Heidelberg, Germany

Printed by Libri Plureos GmbH
in Hamburg, Germany